Springer Theses

Recognizing Outstanding Ph.D. Research

Aims and Scope

The series "Springer Theses" brings together a selection of the very best Ph.D. theses from around the world and across the physical sciences. Nominated and endorsed by two recognized specialists, each published volume has been selected for its scientific excellence and the high impact of its contents for the pertinent field of research. For greater accessibility to non-specialists, the published versions include an extended introduction, as well as a foreword by the student's supervisor explaining the special relevance of the work for the field. As a whole, the series will provide a valuable resource both for newcomers to the research fields described, and for other scientists seeking detailed background information on special questions. Finally, it provides an accredited documentation of the valuable contributions made by today's younger generation of scientists.

Theses may be nominated for publication in this series by heads of department at internationally leading universities or institutes and should fulfill all of the following criteria

- They must be written in good English.
- The topic should fall within the confines of Chemistry, Physics, Earth Sciences, Engineering and related interdisciplinary fields such as Materials, Nanoscience, Chemical Engineering, Complex Systems and Biophysics.
- The work reported in the thesis must represent a significant scientific advance.
- If the thesis includes previously published material, permission to reproduce this must be gained from the respective copyright holder (a maximum 30% of the thesis should be a verbatim reproduction from the author's previous publications).
- They must have been examined and passed during the 12 months prior to nomination.
- Each thesis should include a foreword by the supervisor outlining the significance of its content.
- The theses should have a clearly defined structure including an introduction accessible to new PhD students and scientists not expert in the relevant field.

Indexed by zbMATH.

More information about this series at http://www.springer.com/series/8790

Rina Tazai

Theory of Multipole Fluctuation Mediated Superconductivity and Multipole Phase

Important Roles of Many Body Effects and Strong Spin-Orbit Coupling

Doctoral Thesis accepted by
Nagoya University, Nagoya, Japan

Springer

Author
Dr. Rina Tazai
Department of Physics
Nagoya University
Nagoya, Japan

Supervisor
Prof. Hiroshi Kontani
Department of Physics
Nagoya University
Nagoya, Japan

ISSN 2190-5053 ISSN 2190-5061 (electronic)
Springer Theses
ISBN 978-981-16-1025-7 ISBN 978-981-16-1026-4 (eBook)
https://doi.org/10.1007/978-981-16-1026-4

This Springer imprint is published by the registered company Springer Nature Singapore Pte Ltd.
The registered company address is: 152 Beach Road, #21-01/04 Gateway East, Singapore 189721, Singapore

To my family, mother, father sister and grand parents, who support my life.

To my supervisor, Prof. Hiroshi Kontani. Without his sincere support, it would not be possible to publish the Ph.D. thesis.

To my friends who always cheer me on.

Supervisor's Foreword

Strong electron correlations lead to various exotic electronic states in many transition metals and f-electron compounds. Famous examples are the unconventional non-BCS superconductivity and the higher order multiple orders, *e.g.*, quadrupole and octupole orders. To understand these exotic phenomena, one have to develop the theory of strongly correlated electron systems by going beyond the mean-field-level approximations. This is an important issue in condensed matter theory.

Recently, in Fe-based superconductors, cuprate superconductors, nematic order due to electric quadrupole order (like orbital order and bond order) has been discovered one after another. This important discovery has triggered the progress of the theoretical methods for $3d$ electron systems. In f-electron system, however, coexistence of strong Coulomb interaction U and strong spin-orbit interaction (SOI) hindered the progress of the theory.

Dr. Rina Tazai has been engaged in theoretical research of strongly correlated electing systems, by focusing on the higher order correlation effect called the "vertex corrections (VC)". She first applied the functional renormalization group (fRG) method to a multi-orbital electron system, and discovered the "paramagnon interference mechanism" of the orbital order that is described by the VC. Furthermore, it was shown that the orbital fluctuation mediates strong attractive force of the Cooper pair.

Also, Dr. Tazai has studied the f-electron systems, in which strong U and strong SOI coexist, based on the Feynman diagram field theory. In these systems, higher multiple orders are active, thanks to the strong SOI of f-ion. She introduced the electron correlation effect into the multipolar systems, and revealed that various multipole fluctuations develop cooperatively, thanks to the interference mechanism among different fluctuations. In particular, it was found that the electric hexadecapole fluctuations mediate the s-wave superconductivity without sign reversal, against the strong U between f-electrons. This theory was fruitfully applied to a heavy fermion superconductor $CeCu_2Si_2$, which was recently revealed as a s-wave superconductor by impurity effect and penetration depth measurements.

Furthermore, Dr. Tazai has applied the multipole interference theory to rich multipole order physics in heavy fermion systems. She studied a typical multipole compound CeB_6, and succeeded in explaining the correct quadruple order and the magnetic-field-induced octupole order in terms of the itinerant Fermi liquid picture.

In summary, Dr. Tazai has developed an original theory of superconductivity and multipole order physics in f-electron systems with multipolar degrees of freedom. She discovered the important role of the quantum interference effects in f-electron multipolar systems, by performing the detailed analyses based on the fRG and the Feynman diagram field theory. This theory has been applied to the s-wave superconductivity in $CeCu_2Si_2$ and the quadruple order in CeB_5 successfully, and it will be fruitfully applied to other heavy fermion compounds in future studies.

Nagoya, Japan
November 2020

Hiroshi Kontani

Preface

This book is summary of my Ph.D. thesis published in Department of Science, Nagoya University in March 2020. The main topic is many body effects and spin-orbit coupling in strongly correlated electron systems (SCES). Until now, various exotic quantum phenomena have been reported in SCES such as phase transition with nontrivial order parameters and transport phenomena. However, their microscopic origin is still long-standing problem since it requires to solve many body problem among a number of electrons $N \sim 10^{23}$. Thus, to build a new microscopic theory for explaining the exotic phenomena is one of the central issues in SCES.

In addition to the many body effect, spin-orbit coupling (SOI) is also one of the unsolved topics in SCES. The most prominent effects of SOI are hybridization of spin and real space. In this case, multipole degrees of freedom is activated. Various interesting phenomena such as hidden order, anomalous transport, and unconventional superconductivity appear due to the multipole degrees of freedom.

In the present study, we aim to understand these interesting phenomena on an equal footing. Especially, we focus on the microscopic origin of "unconventional superconductivity" and "hidden order" motivated by recent progresses in the experimental field. This book contains five chapters, which contain different topics.

In Chap. 1, we review microscopic theory of superconductivity and introduce the concept of multipole degrees of freedom.

In Chap. 2, we investigate roles of many body effects in multi-orbital superconductor based on functional renormalization group (fRG).

In Chap. 3, we study a new microscopic theory of superconductivity, in which anti-ferromagnetic fluctuations and electron-phonon interaction cooperatively enhance the T_C.

In Chap. 4, we study superconducting mechanism of fully gapped superconducting phase in $CeCu_2Si_2$.

In Chap. 5, we study microscopic origin of hidden order in CeB_6 considering the quantum interference among magnetic multipole fluctuations.

Nagoya, Japan
November 2020

Rina Tazai

Parts of this thesis have been published in the following journal articles:

1. R. Tazai, and H. Kontani, "Multipole Fluctuation theory for heavy fermion systems: Application to multipole orders in CeB6", Physical Review B; rapid communication, vol. **100**, 241103 (R) (2019).
2. R. Tazai, and H. Kontani, "Hexadecapole fluctuation mechanism for s-wave heavy fermion superconductor CeCu2Si2: Interplay between intra- and inter-orbital Cooper pairs", J. Phys. Soc. Jpn, vol. **88**, 6, 063701 (2019).
3. R. Tazai, and H. Kontani, "Fully gapped s-wave Superconductivity enhanced by magnetic criticality in heavy fermion system", Phys. Rev. B, vol. **98**, 205107 (2018). Editor's suggestion
4. R. Tazai, Y. Yamakawa, M. Tsuchiizu, and H. Kontani, "Plain s-wave Superconductivity near the magnetic criticality: Enhancement of attractive electron-boson coupling vertex corrections", J. Phys. Soc. Jpn, vol. **86**, 073703 (2017).
5. R. Tazai, Y. Yamakawa, M. Tsuchiizu, and H. Kontani, "Functional-Renormalization Group Study for the Orbital-Fluctuation-Mediated Superconductivity", Phys. Rev. B, vol. **94**, 115155 (2016).

Acknowledgements

I would like to thank all related members of Sc laboratory, Dr. Onari, Dr. Kobayashi, Dr. Tsuchiizu, Dr. Yamakawa, and colleagues. Their interesting discussion with sophisticated knowledge really inspired me. Every day in the laboratory was fulfilling for me. Especially, I would like to express sincere gratitude to my supervisor, Prof. Hiroshi Kontani. I learned a lot from him through my graduate education. He taught me how to study physics, write a paper, and give a presentation. My sincere thanks also go to everybody who supported my study in every situation. Their kind advices improved my research.

Finally, I would like to express my deepest thanks to my family (my parents, younger sister, and my dog) for supporting my life for a long time.

This Ph.D. thesis has been supported by leading program in Green Natural Science in Nagoya University and KAKENHI (No.18J12852) from the Japan Society for the Promotion of Science (JSPS).

Contents

Acronyms

AFM	Anti-ferromagnetic
AL	Aslamazov-Larkin
BCS	Bardeen-Cooper-Schrieffer
CEF	Crystalline Electric Field
cRPA	constrained Random Phase Approximation
DoS	Density of States
el-ph	electron-phonon
FLEX	FLuctuation EXchange
FM	Ferromagnetic
fRG	functional Renormalization Group
HF	Heavy Fermion
ME	Migdal-Eliashberg
MT	Matki-Thompson
OP	Order Parameter
PAM	Periodic Anderson Model
p-h, ph	particle-hole
p-p, pp	particle-particle
QCP	Quantum Critical Point
RG	Renormalization Group
RPA	Random Phase Approximation
SC	Superconducting
SCES	Strongly Correlated Electron System
SC-VC	Self-Consistent Vertex Correction
SOI	Spin-Orbit Interaction
SSC	Singlet Superconducting
TSC	Triplet Superconducting
VC	Vertex Correction
U-VC	Vertex Correction for Coulomb interaction U
χ-VC	Vertex Correction for susceptibility χ

Chapter 1
Common Introduction

1.1 Superconductivity in SCES

In the strongly correlated electron systems (SCES), there remains considerable unsolved problems. We review some of them as an introduction to our present study. Until now, various interesting quantum phenomena have been observed in SCES. For instance, phase transition with nontrivial order parameters and transport phenomena have been studied intensively. However, it is still long-standing problem to understand these phenomena microscopically, since it requires to solve "many body problem" among a number of electrons $N \sim 10^{23}$. Therefore, development of microscopic many body theory is one of the important issues in SCES. In this study, we aim to understand these phenomena in the same framework. Especially, we focus on microscopic origin of "unconventional superconductivity" and "hidden order" motivated by recent progress in the experimental field.

First, we review a brief history of theoretical studies on microscopic origin of superconductivity. In 1957, Bardeen, Cooper, and Schrieffer discovered that phonon-mediated retarded attraction works as a glue for Cooper pairs [1]. Now, it is well known as BCS theory and succeeded in explaining the characteristic features of weakly correlated superconductor mainly composed of s- and p-orbital electrons.

On the other hand, the first discovery of superconductor in SCES was reported in heavy fermion (HF) system $CeCu_2Si_2$ in 1979. After that, various transition metals such as cuprates, ruthenate, and iron-based compounds were also recorded. However, the microscopic origin of these superconductors could not be understood within BCS theory since various kinds of particle-hole (p-h) instabilities due to the strong Coulomb repulsion can overwhelm BCS-like electron-phonon (el-ph) interaction. In this case, non-BCS superconducting (SC) states may appear. In facts, various p-h-ordered phases such as magnetic, charge, and orbital-ordered phase appear near SC phase in SCES. Thus, new theory of SC paring mechanism beyond BCS formalism is required for understanding SCES.

R. Tazai, *Theory of Multipole Fluctuation Mediated Superconductivity and Multipole Phase*, Springer Theses,
https://doi.org/10.1007/978-981-16-1026-4_1

To solve the problem, simple approximation based on random phase approximation + Migdal-Eliashberg theorem (RPA+ME) has been used for years. It succeeded in explaining unconventional SC phase near anti-ferromagnetic (AFM) phase in SCES, such as cuprates superconductors. The essence of RPA+ME theory is listed as follows:

1. p-h excitation is described by simple one-loop process in Fig. 1.1b.
2. p-h fluctuations are given by p-h ladder and p-h bubble processes.
3. p-p instability in forming Cooper pairs is given by p-p ladder process.
4. Coupling constant between electron and p-h fluctuations is bare U.

Here, 1 and 2 correspond to RPA theory while 3 and 4 represent ME theory. Especially, 4 is often called Migdal approximation [3]. Based on 1 and 2, development of AFM fluctuations is explained, and AFM fluctuations work as repulsion for Cooper pairs according to 3 and 4. Therefore, one of the main conclusions from RPA+ME theory is that nodal SC pairing state emerges to gain condensation energy [2]. It is well known that many superconductors in SCES show similar phase diagram, which means that AFM phase is close to SC phase as shown in Fig. 1.1a. However, recent improved experiments revealed isotropic s-wave SC states emerge even near AFM phase in SCES, such as Fe-compounds, organic system, and HF system $CeCu_2Si_2$. In particular, discovery of s-wave SC state in $CeCu_2Si_2$ in 2017 surprised researchers since it had been believed as d-wave superconductor for a long time after 1979. These discoveries revealed that RPA+ME theory fails in many systems and unreliable. Therefore, we have to go beyond conventional RPA+ME theory.

In addition, SC phase associated with multi-orbital nature has been studied intensively. For example, SC phase emerges near the orbital-ordered phase in Fe-based and f-electron superconductors. Therefore, electric fluctuations due to the orbital degrees of freedom can be important in multi-orbital superconductors. However, in the theoretical way, it is difficult to explain the emergence of orbital ordering/fluctuations within the realistic condition. In fact, only p-h fluctuations related to the spin degrees

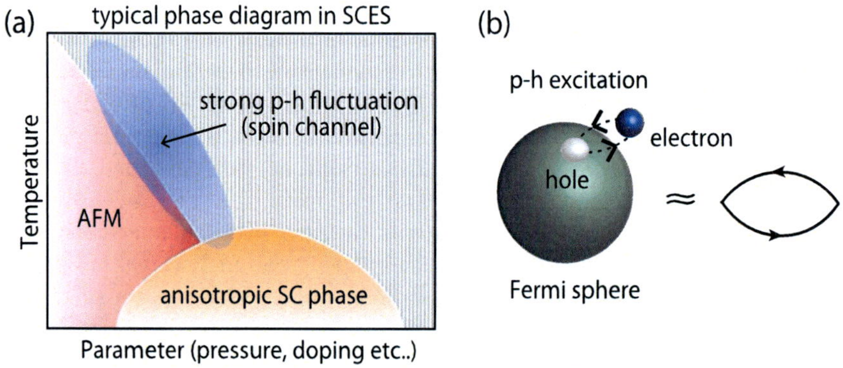

Fig. 1.1 **a** Typical phase diagram often seen in SCES, such as cuprates, Fe-based and HF compounds. **b** p-h excitation due to one-loop process

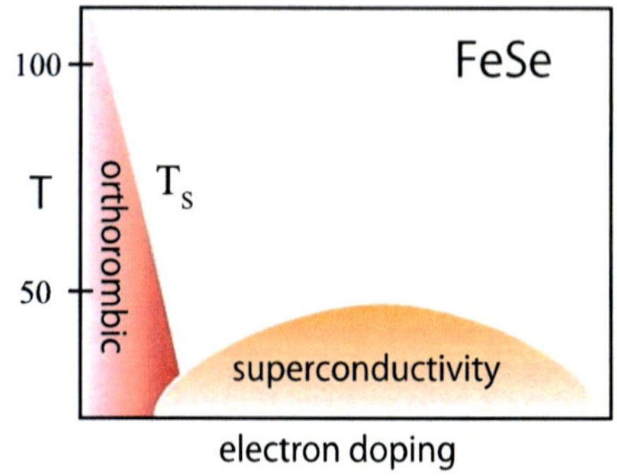

Fig. 1.2 Schematic phase diagram of FeSe. Structural phase transition appears without the magnetic ordering

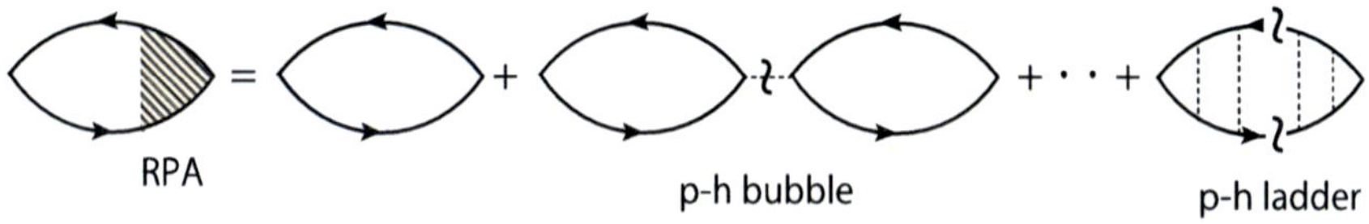

Fig. 1.3 p-h fluctuations due to p-h bubble and ladder processes. Solid and dotted lines correspond to electron propagator and Coulomb repulsion, respectively

of freedom become dominant over that of orbital channel within the conventional mean-field scheme, such as RPA and fluctuation exchange (FLEX) approximation [4] (Fig. 1.2).

Hereafter, we explain significant roles of many body effects beyond conventional theory. Before that, we review conventional RPA+ME theory in more detail. In Fig. 1.3, we show the Feynman diagram of p-h fluctuations by RPA. Solid lines represent one-particle propagator (Green function) and dotted line corresponds to electron-electron interaction, such as Coulomb interaction U. Filled region shows reducible three-point vertex function. We consider simple one-loop p-h excitation corresponding to the first term of right-hand side (rhs) of the figure within RPA. Then, the Coulomb interaction U is considered by bubble and ladder process as shown in the second and third terms of rhs in Fig. 1.3, respectively. Then, p-p instability is obtained by p-p ladder process as shown in Fig. 1.4. The red (black)-colored rectangle shows reducible (irreducible) four-point vertex function of p-p channel. p-p channel four-point vertex is composed of p-h fluctuations. In ME approximation, coupling constants between Green function and p-h fluctuations are given by the bare Coulomb interaction U represented as the right diagram in Fig. 1.4.

On the other hand, in the present study, we consider many body effects beyond RPA+ME theory. In this case, irreducible p-h fluctuations are dressed by three-point vertex correction (VC) as shown in Fig. 1.2. The red-colored region represents the three-point VC due to the many body effects beyond RPA. Here, three-point VC is

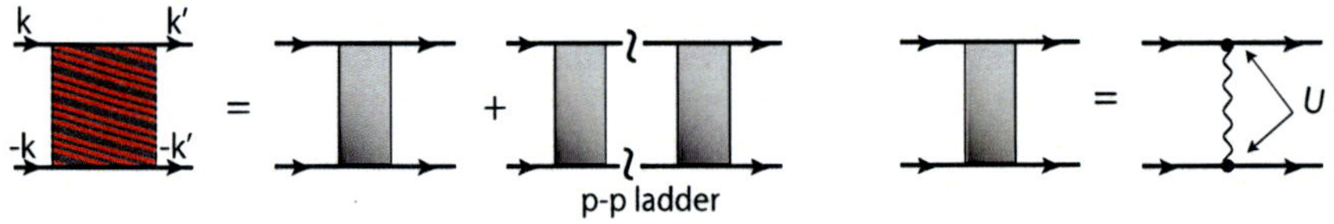

Fig. 1.4 SC paring instability due to p-p ladder process. Red (black) filled rectangular shows reducible (irreducible) four-point vertex function, respectively. Wavy line corresponds to p-h fluctuations

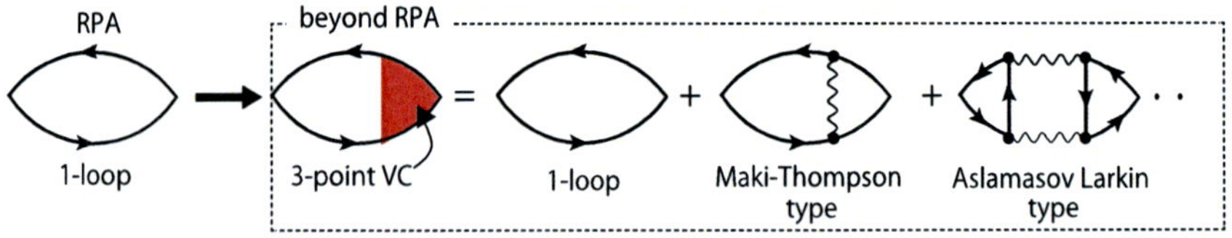

Fig. 1.5 p-h fluctuations beyond RPA+ME theory. The red-colored triangle shows three-point VC

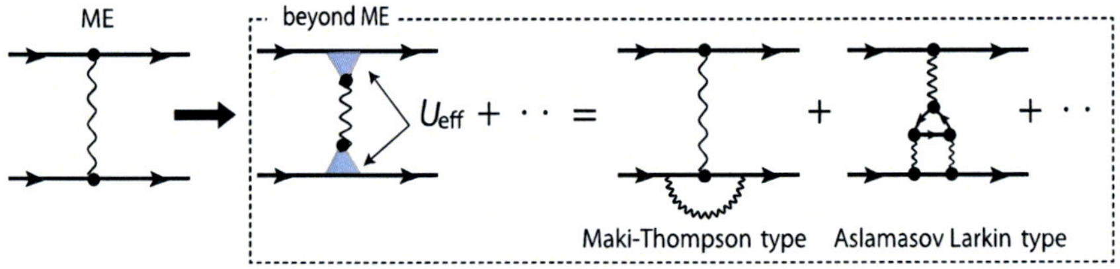

Fig. 1.6 Blue-colored triangle shows three-point VC beyond ME. The VC due to the first (second) order of p-h fluctuations corresponds to MT (AL) term

reducible with respect to U. The first term of the rhs of the figure shows one-loop process included in RPA. In contrast, the second and third terms are VC neglected in RPA. Through our study, we found that these VCs play significant roles for SC paring as well as hidden ordering. Thus, we call them χ-VC. In more detail, the second term is VC due to the single p-h fluctuation, which is called Maki-Thompson (MT) term. The third one corresponds to second order of p-h fluctuation, which is called Aslamazov-Larkin (AL) term (Fig. 1.5).

Figure 1.6 represents the irreducible four-point vertex in p-p ladder beyond ME theory. The blue-colored triangle gives three-point VC neglected in ME theory. We call the VC in p-p ladder U-VC. The U-VC also includes AL- and MT-type diagrams. In the present study, we revealed that both U-VC and χ-VC play significant roles to understand the SC paring mechanism in SCES.

In general, it is difficult to understand what kind of scattering process is essentially important in SCES, since it require us to consider all of the possible processes including higher order terms than abovementioned processes. To over come this difficulty, we study fRG+constrained RPA (cRPA) in the following section. In this method, parquet-type scatterings are automatically generated only by solving differential equation. Therefore, obtained results are unbiased and quite reliable. However, it is still hard to identify the essential process solely by fRG+cRPA, since a number of minor processes are automatically included. Thus, it is important to analyze by using another method, such as diagrammatic calculation. For this reason, we need to compare the theoretical results by using both fRG and diagrammatic calculation.

Now, we are at the starting point for constructing a new microscopic theory of SC paring in SCES. Hereafter, we move into our present study. To go beyond RPA+ME theory, we focus on higher order many body effects. Through our study in multi-orbital SCES, we found that violation of abovementioned 1 and 4 occurs and they are replaced with

1. Many body effects beyond one-loop process are important for p-h fluctuations.
2. Coupling constant changes from bare U to dressed one by many body effect.

These findings are our main results in the present book. Thus, we show more detailed explanation on this point in each section. Based on the results, we reconsider following questions:

- Q1: What kind of scattering process is significant in multi-orbital SCES?
- Q2: Is it true that AFM fluctuations and el-ph coupling always compete in SC paring mechanism as predicted in RPA+ME theory?
- Q3: How can we understand s-wave SC state in $CeCu_2Si_2$ with large U?

In addition, we notice that our findings is also helpful to understand long-standing issues in normal states as well as SC states. For instance, microscopic origin of hidden-ordered phase in HF system remains significant problem. Therefore, we study on CeB_6 as one of the typical hidden-ordered systems. Then, we add the following question:

Q4: How can we understand hidden-ordered phase by many body effects?

To answer these fundamental questions, we study SC paring and hidden ordering mechanism in SCES based on the common "many body effects mechanism".

To understand the questions Q3 and Q4, strong spin-orbit interaction (SOI) has to be taken into account. In this case, f-electrons are characterized by total angular momentum J instead of spin and orbital angular momentum S, L. Therefore, p-h fluctuations are defined in J-space, which we call multipole fluctuation. Concept of multipole fluctuation is introduced in Chap. 4. Development of higher rank multipole fluctuations is characteristic property of f-electron systems and brings various interesting phenomena. For this reason, the fifth question is

Q5: How do multipole fluctuations work in HF system with many body effects?

In this paper, we discuss abovementioned questions in the following four sections (Chaps. 2–5). In Chap. 2, we study d-orbital Hubbard model based on functional-renormalized-group+constrained RPA (fRG+cRPA) method to discover fundamental properties of many body effect beyond RPA+ME, corresponding to Q1. In Chap. 3, we study Hubbard-Holstein model with phonon-mediated interaction to answer the Q2. After that, in Chap. 4, we move into HF superconductor $CeCu_2Si_2$ with strong

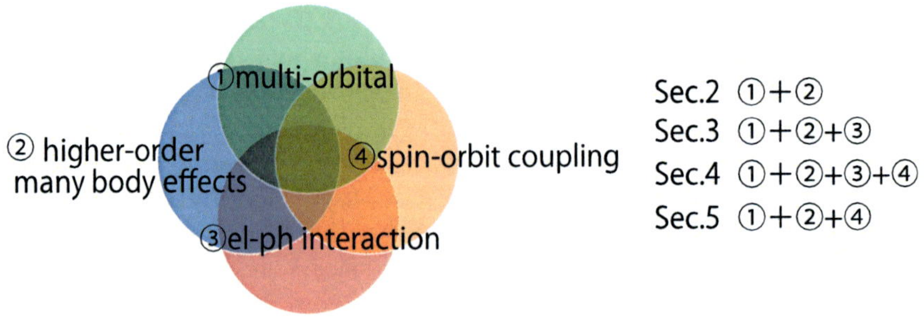

Fig. 1.7 Map of the present study. We propose a microscopic origin of SC paring and hidden ordering considering higher order many body effects, el-ph interaction, SOI, and multi-orbital nature

SOI to answer Q3 and Q5. In Chap. 5, we propose a microscopic origin of hidden-ordered phase in CeB_6 related to Q4 and Q5. We summarized our present study in the map of Fig. 1.7. After all, we obtain the following results, which answer the questions Q1∼Q5:

1. Mode coupling effect between orbital and spin degrees of freedom plays essential roles for SC paring in multi-orbital SCES system.
2. AFM fluctuations and phonon-mediated interaction can work cooperatively and enhance the T_c of isotropic SC paring.
3. Mode coupling effect is significant in f-electrons due to the strong SOI. Interference between electric and magnetic multipole fluctuations stabilizes s-wave SC.
4. Interference among different rank multipole fluctuations causes the hidden-ordered phase.

1.2 Multipole Degrees of Freedom

In this section, we introduce "multipole" to classify the p-h order parameters in f-electron system with strong SOI. Before that, we demonstrate how to define the multipole operator (OP) based on d-electron model for simplicity. First, we consider d_{zx}- and d_{yz}-orbital wave function given by

$$\psi_{zx}(\mathbf{r}) = R_{n,2}(r)\frac{1}{\sqrt{2}}\{Y_{2,1}(\theta,\phi) + Y_{2,-1}(\theta,\phi)\}, \tag{1.1}$$

$$\psi_{yz}(\mathbf{r}) = R_{n,2}(r)\frac{i}{\sqrt{2}}\{Y_{2,1}(\theta,\phi) - Y_{2,-1}(\theta,\phi)\}, \tag{1.2}$$

where $R_{n,l}(r)$ and $Y_{l,m}(\theta, \phi)$ are radial distribution function and spherical harmonics, respectively. The index (n, l, m) denotes (principal, azimuth, magnetic) quantum number. By using the wave function, the possible p-h order parameter (OP) is expressed as

$$OP = \sum_{a,a'=dzx,dyz} \int d\mathbf{r}_i \psi_a^*(\mathbf{r}_i) f_{aa'}(\mathbf{r}_i) \psi_{a'}(\mathbf{r}_i) \tag{1.3}$$

where i denotes the site index. The function $f_{aa'}(\mathbf{r}_i)$ represents the anisotropy of electrons charge distribution and defines the symmetry of the local order parameter. Here, the order parameter function f is also written by using $R_{n,l}(r)$ and $Y_{l,m}(\theta, \phi)$. Therefore, the OP is expressed as

$$OP \propto \int d\phi Y^*_{l,m1} Y_{l',m'} Y_{l,m2} = \delta(m1 - m' - m2) \quad \left(f_{aa'}(\mathbf{r}_i) \propto Y_{l',m'}\right). \tag{1.4}$$

In this case, possible combinations for $(l'm')$ in $Y_{l',m'}(\theta, \phi)$ are nine patterns given by

$$f_{aa'}(\mathbf{r}_i) \propto Y_{0,0}, Y_{1,0}, Y_{1,\pm1}, Y_{2,0}, Y_{2,\pm1}, Y_{2,\pm2}, \tag{1.5}$$

These nine possible order parameters are classified into multipole channel by m'. $f_{aa'}(\mathbf{r}_i)$ for $m' = (0, 1, 2)$ corresponds to (monopole, dipole, quadrupole), which belongs to rank (0, 1, 2). In more general, when the wave functions of the ground states are written by $Y_{l,m=a\sim b}$, the possible $f_{aa'}(\mathbf{r}_i)$ is given by

$$f_{aa'}(\mathbf{r}_i) \propto Y_{l',0}, Y_{l',\pm1} \sim Y_{l',\pm(b-a)} \quad (0 \le l' \le l). \tag{1.6}$$

Therefore, if we consider all of the $t_{2g} + e_g$-orbital, which means that the wave function is written by $Y_{2,0} Y_{2,\pm1}, Y_{2,\pm2}$, we obtain 25-type possible orders up to the hexadecapole (rank 4).

On the other hand, the "active" multipole orders are only four types (nine types) in the case of d_{zx}- and d_{yz}-orbital ($t_{2g} + e_g$-orbital) system since the independent degrees of freedom are given by square of the orbital number. For instance, in the monopole case ($f_{aa'}(\mathbf{r}_i) \propto 1$), the OP is rewritten as

$$\begin{aligned} OP(\text{mono}) &= \sum_{a,a'=dzx,dyz} \int d\mathbf{r}_i \psi_a^*(\mathbf{r}_i) \sqrt{x^2+y^2+z^2} \psi_{a'}(\mathbf{r}_i) \\ &= \int d\mathbf{r}_i \sqrt{x^2+y^2+z^2} (\psi^*_{dzx}(\mathbf{r}_i)\psi_{dzx}(\mathbf{r}_i) + \psi^*_{dyz}(\mathbf{r}_i)\psi_{dyz}(\mathbf{r}_i)) \propto n_{zx} + n_{yz}. \end{aligned} \tag{1.7}$$

To derive the final expression in Eq. (1.7), we use $\int d\mathbf{r}_i \sqrt{x^2+y^2+z^2} \psi^*_{dzx} \psi_{dyz} = 0$ since integrands are odd function for x or y. Therefore, the order parameter corresponds to $n_{zx} + n_{yz}$. Moreover, we obtain the following matrix representation focusing on the orbital dependence of $\hat{f}(\mathbf{r}_i)$ for monopole.

$$OP(\text{mono}) = \begin{array}{c} \\ \psi^*_{zx} \\ \psi^*_{yz} \end{array} \begin{array}{c} \begin{array}{cc} \psi_{zx} & \psi_{yz} \end{array} \\ \begin{pmatrix} 1 & 0 \\ 0 & 1 \end{pmatrix} \end{array}. \tag{1.8}$$

In the same way, we also obtain the $x^2 - y^2$-type quadrupole OP given by

$$\begin{aligned} OP(\text{quad} : x^2 - y^2) &= \sum_{a,a'=dxz,dyz} \int d\mathbf{r}_i \psi^*_a(\mathbf{r}_i)(x^2 - y^2)\psi_{a'}(\mathbf{r}_i) \\ &= \int d\mathbf{r}_i (x^2 - y^2)(\psi^*_{dzx}(\mathbf{r}_i)\psi_{dzx}(\mathbf{r}_i) - \psi^*_{dyz}(\mathbf{r}_i)\psi_{dyz}(\mathbf{r}_i)). \end{aligned} \tag{1.9}$$

In this case, the order parameter is given by $\left\langle \psi^*_{dzx}(\mathbf{r}_i)\psi_{dzx}(\mathbf{r}_i) - \psi^*_{dyz}(\mathbf{r}_i)\psi_{dyz}(\mathbf{r}_i) \right\rangle$ and

$$OP(\text{quad} : x^2 - y^2) = \begin{array}{c} \\ \psi^*_{zx} \\ \psi^*_{yz} \end{array} \begin{array}{c} \begin{array}{cc} \psi_{zx} & \psi_{yz} \end{array} \\ \begin{pmatrix} 1 & 0 \\ 0 & -1 \end{pmatrix} \end{array} \propto \hat{L}_x^2 - \hat{L}_y^2. \tag{1.10}$$

In addition, xy-type quadrupole OP is also independent OP to abovementioned two OPs.

$$OP(\text{quad} : xy) = \int d\mathbf{r}_i xy(\psi^*_{dzx}(\mathbf{r}_i)\psi_{dyz}(\mathbf{r}_i) + \psi^*_{dyz}(\mathbf{r}_i)\psi_{dzx}(\mathbf{r}_i)), \tag{1.11}$$

where the order parameter corresponds to $\left\langle \psi^*_{dzx}(\mathbf{r}_i)\psi_{dyz}(\mathbf{r}_i) + \psi^*_{dyz}(\mathbf{r}_i)\psi_{dzx}(\mathbf{r}_i) \right\rangle$

$$OP(\text{quad} : xy) = \begin{array}{c} \\ \psi^*_{zx} \\ \psi^*_{yz} \end{array} \begin{array}{c} \begin{array}{cc} \psi_{zx} & \psi_{yz} \end{array} \\ \begin{pmatrix} 0 & 1 \\ 1 & 0 \end{pmatrix} \end{array} \propto \hat{L}_x \hat{L}_y. \tag{1.12}$$

We note that xz- or yz-type quadrupole OP goes to zero due to the symmetry of the wave function. The $3z^2 - r^2$-type quadrupole OP is not independent OP since we obtain the following equation:

$$\begin{aligned} OP(\text{quad} : 3z^2 - r^2) &= \sum_{a,a'=dxz,dyz} \int d\mathbf{r}_i \psi^*_a(\mathbf{r}_i)(3z^2 - r^2)\psi_{a'}(\mathbf{r}_i) \\ &= \int d\mathbf{r}_i (3z^2 - r^2)(\psi^*_{dzx}(\mathbf{r}_i)\psi_{dzx}(\mathbf{r}_i) + \psi^*_{dyz}(\mathbf{r}_i)\psi_{dyz}(\mathbf{r}_i)), \end{aligned} \tag{1.13}$$

Table 1.1 Active multipole in the present model

	d_{zx}, d_{yz}	$t_{2g} + e_g$	Γ_7	Γ_8
Number of the active OPs	4	25	16	16
The highest rank	Quadrupole	Hexadecapole	Dotriacontapole	Octupole
Section	2, 3	–	4	5

where the order parameter corresponds to $n_{zx} + n_{yz}$, which is the same as that of monopole. Now, we have already obtained three multipole OP including one monopole and two quadrupole OPs. The final one is dipole one, which is given by

$$OP(\text{dipole}) = \begin{matrix} & \psi_{zx} & \psi_{yz} \\ \psi^*_{zx} & 0 & i \\ \psi^*_{yz} & -i & 0 \end{matrix} \propto \hat{L}_z. \tag{1.14}$$

In the dipole-ordered phase, time reversal symmetry is broken even in the orbital space. For instance, local circular current emerges in the dipole phase. As we show in this section, the OP matrix is written by using the orbital angular momentum operator $\hat{L}$. On the contrary, with the strong SOI, the $\hat{L}$ changes to $\hat{J} = \hat{L} + \hat{S}$. Therefore, it is naturally expected that the rank of OP with SOI tends to be higher than that without the SOI. Here, the active OPs in multipole basis are summarized in Table 1.1.

References

1. J. Bardeen, L.N. Cooper, J.R. Schrieffer, Phys. Rev. **108**, 1175 (1957)
2. J.R. Schrieffer, J. Low Temp. Phys **99**, 314 (1995)
3. A.B. Migdal, J. Exptl. Thoret. Phys. **34**, 1438 (1958)
4. N.E. Bickers, S.R. White, Phys. Rev. B **43**, 8044 (1991)

Chapter 2
Functional Renormalization Group (fRG) Study

2.1 Background

Spin Triplet Superconductivity

As well known, more than 99% of discovered superconductor show spin singlet Cooper (SSC) pairing states. The wave function of the spin singlet pairing state is written by

$$\Psi_{SSC} \propto |\uparrow\downarrow\rangle - |\downarrow\uparrow\rangle, \tag{2.1}$$

where the total spin momentum is $S = 0$. On the other hand, spin triplet superconductivity (TSC) was reported by some experiments. The schematic picture of triplet Cooper pair is shown in Fig. 2.1. The TSC pair wave function is given by

$$\Psi_{TSC} \propto |\uparrow\downarrow\rangle + |\downarrow\uparrow\rangle, |\uparrow\uparrow\rangle, |\downarrow\downarrow\rangle, \tag{2.2}$$

where the total spin becomes $S = 1$. Thus, the difference of triplet and singlet SC state appears in the spin degrees of freedom. For instance, Knight shift by NMR does not change under the TSC transition, while it is drastically suppressed in the SSC states. In addition, the parity of the pair wave function is different between TSC and SSC. in the case of even-frequency and zero-momentum SC pairing states. The momentum dependence of pairing wave function is

$$\begin{aligned}\Psi_{SSC} &\propto |\boldsymbol{k}, -\boldsymbol{k}\rangle + |-\boldsymbol{k}, \boldsymbol{k}\rangle \propto Y_{l,m} \ (l = 0, 2, 4, 6..) \\ \Psi_{TSC} &\propto |\boldsymbol{k}, -\boldsymbol{k}\rangle - |-\boldsymbol{k}, \boldsymbol{k}\rangle \propto Y_{l,m} \ (l = 1, 3, 5, 7..).\end{aligned} \tag{2.3}$$

Therefore, s-wave ($l = 0$) and d-wave ($l = 2$) SC states appear as SSC states, while p-wave ($l = 0$) and f-wave ($l = 2$) SC states belong to SSC states. Except for s-wave

R. Tazai, *Theory of Multipole Fluctuation Mediated Superconductivity and Multipole Phase*, Springer Theses,
https://doi.org/10.1007/978-981-16-1026-4_2

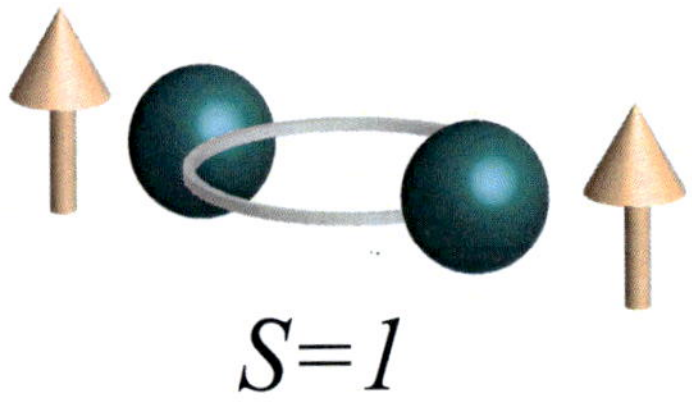

Fig. 2.1 Triplet Cooper pair with total spin $S = 1$

	heat capacity	penetration depth
full gap	$\exp(-\Delta/T)$	$\exp(-\Delta/T)$
point node	T^2	T
line node	T^3	T^2

Fig. 2.2 T-dependence of observable quantity by heat capacity and penetration depth measurement

Table 2.1 Relation between spin/charge/orbital fluctuations and TSC/SSC

	Spin fluctuation	Charge/orbital fluctuation
SSC	Repulsion $\propto 3\chi^s$	Attraction $\propto \chi^c$
TSC	Attraction $\propto \chi^s$	Attraction $\propto \chi^c$

SC, the gap function has nodes on the Fermi surface, which is detected by some experiments such as heat capacity and penetration depth summarized in Fig. 2.2. Until now, UPt_3, UTe_2, and Sr_2RuO_4 are reported as the TSC metals [1–10] These materials have d- or f-electrons with orbital degeneracy. Therefore, it is naturally expected that the TSC is stabilized by both multi-orbital nature and electric correlation.

However, the microscopic origin of the TSC is still unsolved. Roughly speaking, there are three possibilities for the microscopic origin of TSC:

1. TSC is explained by large ferromagnetic (FM) fluctuation ($Q = 0$).
2. TSC is explained by Ising-like anisotropy of the FM/AFM fluctuation.
3. TSC is explained by interference between orbital fluctuations and AFM fluctuations.

The first possibility is FM fluctuation scenario, which succeeded in explaining the superfluid phase in ^{3}He. The general pairing interaction of SSC and TSC is given by

$$V_{SSC}(\boldsymbol{k}, \boldsymbol{k}') = \frac{3}{2}U^2\chi^s(\boldsymbol{k}-\boldsymbol{k}') - \frac{1}{2}U^2\chi^c(\boldsymbol{k}-\boldsymbol{k}'),$$
$$V_{TSC}(\boldsymbol{k}, \boldsymbol{k}') = -\frac{1}{2}U^2\chi^s(\boldsymbol{k}-\boldsymbol{k}') - \frac{1}{2}U^2\chi^c(\boldsymbol{k}-\boldsymbol{k}'), \quad (2.4)$$

where the repulsion/attraction for Cooper pairing is represented by positive/negative sign. Accordingly, magnetic fluctuations driven SSC phase easily appears due to the factor "3" before $\chi^s(\boldsymbol{k}-\boldsymbol{k}')$ compared with TSC. The relation between the pairing interaction and spin/charge/orbital fluctuations are summarized in the Table 2.1. In the presence of large FM fluctuations, the interaction is simply putted as $\chi^s(\boldsymbol{k}-\boldsymbol{k}') = \chi^s(0)\delta_{\boldsymbol{k},\boldsymbol{k}'}$. Thus, the linearized SC gap equation is

$$\Delta_{SSC}(\boldsymbol{k}) \propto -\frac{3}{2}U^2\chi^s(0)\Delta_{SSC}(\boldsymbol{k}),$$
$$\Delta_{TSC}(\boldsymbol{k}) \propto \frac{1}{2}U^2\chi^s(0)\Delta_{TSC}(\boldsymbol{k}). \tag{2.5}$$

The important difference appears in the sign of the r.h.s. of the gap equations. The positive sign for $\Delta_{TSC}(\boldsymbol{k})$ shows that the TSC gap is stabilized due to the FM fluctuations while the $\Delta_{SSC}(\boldsymbol{k})$ is suppressed due to the negative sign. As a result, large FM fluctuation stabilizes the TSC phase rather than SSC phase.

The second possibility is Ising-like AF fluctuation scenario. In this case, factor $3/2$ of r.h.s. of Eq. (2.4) is reduced to $1/2$ and 1 in the presence of in-plane $\propto S_z \cdot S_z$ and out-of-plane $\propto S_x \cdot S_x$, $S_y \cdot S_y$ spin fluctuation, respectively. In fact, the SC gap equation only due to the in-plane magnetic fluctuation is

$$\Delta_{SSC}(\boldsymbol{k}) \propto -\frac{1}{2}U^2\chi^{sz}(\boldsymbol{k}-\boldsymbol{k}')\Delta_{SSC}(\boldsymbol{k}'),$$
$$\Delta_{TSC}(\boldsymbol{k}) \propto -\frac{1}{2}U^2\chi^{sz}(\boldsymbol{k}-\boldsymbol{k}')\Delta_{TSC}(\boldsymbol{k}'). \tag{2.6}$$

Both the TSC and SSC are stabilized in the same way. Therefore, the Ising-type magnetic fluctuations can cause the TSC phase.

The third possibility is the main topic of our present study [11]. We would like to put the new prediction for the microscopic origin of TSC that

- p-wave TSC is induced by interference of the orbital fluctuations and AFM fluctuations in the absence of the ferro-type or Ising-like spin fluctuations.

Once the orbital fluctuations develop, it is naturally expected that orbital-fluctuation-mediated TSC phase appears since the gap equations due to the orbital fluctuations are written as

$$\Delta_{SSC}(\boldsymbol{k}) \propto \frac{1}{2}U^2\chi^c(\boldsymbol{k}-\boldsymbol{k}')\Delta_{SSC}(\boldsymbol{k}'),$$
$$\Delta_{TSC}(\boldsymbol{k}) \propto \frac{1}{2}U^2\chi^c(\boldsymbol{k}-\boldsymbol{k}')\Delta_{TSC}(\boldsymbol{k}'). \tag{2.7}$$

Thus, the TSC pairing can be induced by orbital fluctuations as well as ordinal SSC pairing. The key idea of the present proposed mechanism is the quantum interference of the orbital and AFM fluctuations, which was dropped in the previous studies. To understand the interference, we have to consider many body effects beyond the conventional Migdal-Elaishberg theory [12]. For this purpose, we use the fRG method introduced in the next section.

2.2 Formalism

In this section, fRG formalisms are summarized. In Sect. 2.2.1, we introduce two-orbital Hubbard model. Based on the Hubbard model, four-point and three-point vertex functions are calculated in Sects. 2.2.2 and 2.2.3, respectively.

2.2.1 *Multi-Orbital Hubbard Model*

Kinetic Term

Here, we introduce two-orbital Hubbard model on 2D square lattice for describing $4d^4$-electron system. Under the tetragonal crystalline electric field (CEF), $4d$-orbitals are splitted into t_{2g} and e_g orbitals as shown in Fig. 2.3a. Then, the ground states become t_{2g} orbitals. In the present study, we only consider d_{xz}- and d_{yz}-orbitals for the simplicity and fix the electron's number at $n = n_{xz} + n_{yz} = 4 \times (2/3) = 2.67$. The Hamiltonian for the kinetic part is written by

$$\hat{H}_0 = \sum_{k\sigma lm} \xi_k^{lm} d^\dagger_{kl\sigma} d_{km\sigma}, \tag{2.8}$$

where l, m takes 1 or 2, which corresponds to d_{xz}- or d_{yz}-orbital, respectively. The energy dispersion is defined as

$$\begin{aligned} \xi_k^{11} &= -2t\cos k_x - 2t^{''}\cos k_y, \\ \xi_k^{22} &= -2t\cos k_y - 2t^{''}\cos k_x, \\ \xi_k^{12} = \xi_k^{21} &= -4t^{'}\sin k_x \sin k_y, \end{aligned} \tag{2.9}$$

where the schematic picture of the hopping is shown in Fig. 2.3b. Energy scale is measured from Fermi energy and the energy unit is $t = 1$. The hopping parameters are fixed at $(t, t^{'}, t^{''}) = (1, 0.1, 0.1)$. The band dispersion and Fermi surfaces of the present model are plotted in Fig. 2.5a, b, respectively. Quasi 1D Fermi surfaces (FS-α and FS-β) are obtained. The nesting vector appears at $\boldsymbol{Q} = (2/3\pi, 2/3\pi)$.

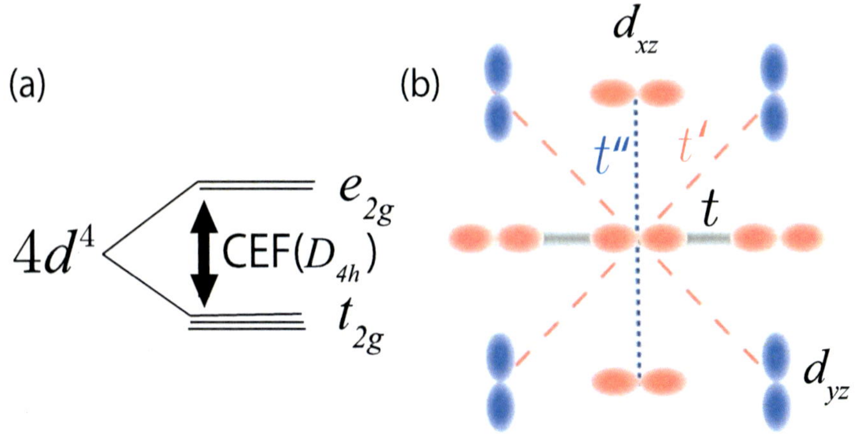

Fig. 2.3 **a** The ground states of the d-electron model. t_{2g}-orbital is filled up with four electrons. **b** The hopping integrals between d_{xz}- and d_{yz}-orbitals

Coulomb Interaction

Here, we introduce the on-site Coulomb interaction, with intra- (inter-)orbital U (U'), pair hopping J' and Hund's coupling J. The definition of U, U', J, J' is given in Fig. 2.4. The Hamiltonian for the interaction part is given by

$$\hat{H}_U = \frac{1}{4}\sum_i \sum_{ll'mm'} \sum_{\sigma\sigma'\rho\rho'} U^{\sigma\sigma'\rho\rho'}_{ll'mm'} d^\dagger_{il\sigma} d_{il'\sigma'} d_{im\rho} d^\dagger_{im'\rho'}, \tag{2.10}$$

$$\hat{U}^{\sigma\sigma'\rho\rho'} = \frac{1}{2}\hat{U}^s \boldsymbol{\sigma}_{\sigma\sigma'} \cdot \boldsymbol{\sigma}_{\rho'\rho} + \frac{1}{2}\hat{U}^c \delta_{\sigma\sigma'}\delta_{\rho'\rho}, \tag{2.11}$$

where $\boldsymbol{\sigma}$ is Pauli matrix vector and $\hat{U}$ is the bare four-point vertex function given by 4×4 tensor, which is anti-symmetrized as $U^{1234}_{5678} = -U^{1324}_{5768}$ based on anti-commutation relation for fermion. The i denotes the site index. All elements in $\hat{U}^{s(c)}$ are listed as follows:

$$\begin{cases} \hat{U}^s = U & (l = l' = m = m') \\ \hat{U}^s = U' & (l = m \neq l' = m') \\ \hat{U}^s = J & (l = l' \neq m = m') \\ \hat{U}^s = J' & (l = m' \neq l' = m), \end{cases} \quad \begin{cases} \hat{U}^c = -U & (l = l' = m = m') \\ \hat{U}^c = U' - 2J & (l = m \neq l' = m') \\ \hat{U}^c = -2U' + J & (l = l' \neq m = m') \\ \hat{U}^c = -J' & (l = m' \neq l' = m). \end{cases} \tag{2.12}$$

Other elements not listed in Eq. (2.12) go to zero. To simplify the model, we assume the relation $J = J' = (U - U')/2$ in the numerical calculation. Therefore, there are only two independent parameters, such as U and J.

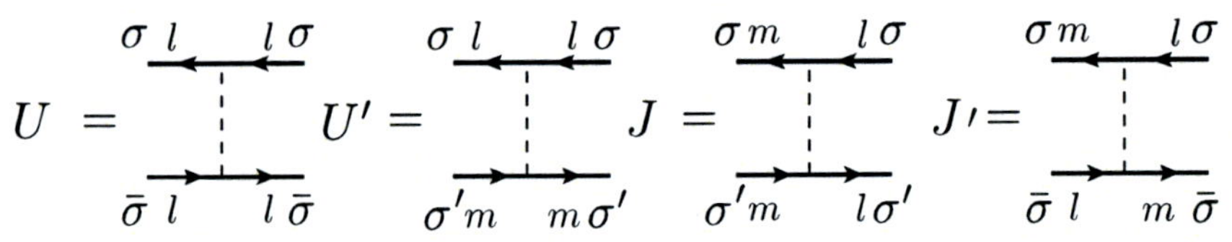

Fig. 2.4 Definition of U, U', J, J' under the condition of $l \neq m$

SU(2) Symmetry

Note that SU(2) symmetry is satisfied in the present Hamiltonian since spin-orbit coupling is absent. In spin conserving system, the interaction Hamiltonian for fermion is given by

$$\hat{H}_U = \sum_{1,2,3,4,\sigma} \Gamma^{\sigma\sigma\sigma\sigma}_{1234} d^\dagger_{1\sigma} d_{2\sigma} d_{3\sigma} d^\dagger_{4\sigma} + \Gamma^{\sigma\bar{\sigma}\sigma\bar{\sigma}}_{1234} d^\dagger_{1\sigma} d_{2\bar{\sigma}} d_{3\sigma} d^\dagger_{4\bar{\sigma}}. \tag{2.13}$$

By using the SU(2) symmetry, the independent component is reduced to only one component. To show that, we perform spin rotation given by

$$\begin{pmatrix} \alpha & -\beta \\ \beta^* & \alpha^* \end{pmatrix} \begin{pmatrix} d_\uparrow \\ d_\downarrow \end{pmatrix} = \begin{pmatrix} d_\uparrow \\ d_\downarrow \end{pmatrix} \quad (|\alpha^2| + |\beta^2| = 1). \tag{2.14}$$

By using the SU(2) condition, in which Hamiltonian does not change under the rotation, we obtain

$$\Gamma^{\sigma\sigma\sigma\sigma}_{1234} - \Gamma^{\sigma\sigma\sigma\sigma}_{1324} = \Gamma^{\sigma\bar{\sigma}\sigma\bar{\sigma}}_{1234} - \Gamma^{\sigma\bar{\sigma}\sigma\bar{\sigma}}_{1324}. \tag{2.15}$$

Therefore, $\Gamma^{\sigma\sigma\sigma\sigma}$ is calculated from $\Gamma^{\sigma\bar{\sigma}\sigma\bar{\sigma}}$ in the presence of SU(2) symmetry.

- We use two-orbital (d_{xz}, d_{yz}) Hubbard model with on-site Coulomb interaction given by four-point vertex function.
- The bare four-point vertex $\hat{U}$ holds SU(2) symmetry.

2.2.2 *Four-Point Vertex Function*

Logarithmic Energy Cutoff

Here, we perform the fRG+constrained RPA (cRPA) [18] to calculate the effective low energy interaction. First, we introduce logarithmic energy cutoff

$$\Lambda_l = \Lambda_0 e^{-l} \quad (l \geq 0). \tag{2.16}$$

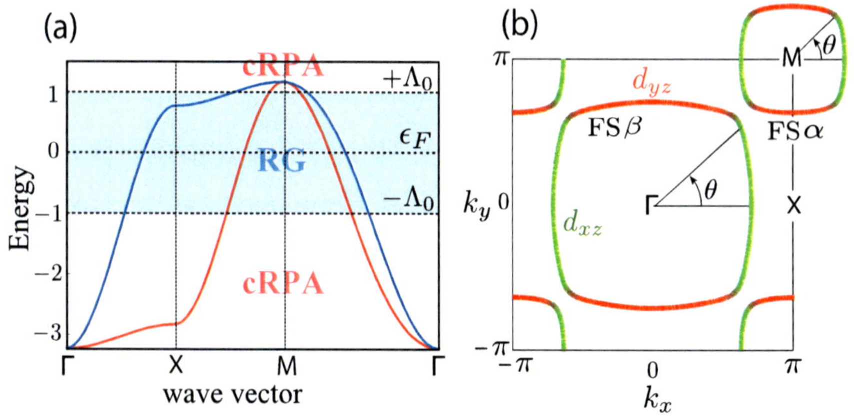

Fig. 2.5 **a** Obtained band dispersion and **b** Fermi surfaces mainly formed by d_{xz} (green) and d_{yz} (red) orbital

Then, d-electrons' energy scale is divided into higher energy ($|\xi_{\boldsymbol{k}}| > \Lambda_0$) and lower energy regions ($|\xi_{\boldsymbol{k}}| < \Lambda_0$). The higher energy scatterings are considered by cRPA with fine $\boldsymbol{k}$ mesh. On the other hand, lower energy regions are calculated by fRG in order to consider higher order many body effects in an unbiased way. The fRG+cRPA concept is justified when higher order many body effects beyond RPA become important only in low-energy region. Based on the present hybrid method, effective low energy interaction is obtained with high accuracy for both energy regions. Here, we set $\Lambda_0 = 1.0$ (< band width) as shown in Fig. 2.5a. It is verified that the present results do not change even in the case of $\Lambda_0 \sim W_{\mathrm{band}}/2$. The RG flow starts from $\Lambda = \Lambda_0$ to the low-energy cutoff $\omega_c \ll T$.

Patch

In exchange for the great advantage of the fRG, taking fine $\boldsymbol{k}$ mesh is difficult in performing the fRG. To overcome this difficulty, we divide lower energy $\boldsymbol{k}$-space into some patches. Figure 2.7a shows the contours for $|\xi_{\boldsymbol{k}}| = \Lambda_0 = 1$ and the center of patches $p_\alpha(p_\beta)$. In the present numerical study, we put 32 patches on each Fermi surfaces. Here are tips of the present fRG+cRPA method:

- High-energy region is considered by cRPA with high accuracy.
- Low-energy region is considered by fRG with higher order many body effects.
- $\boldsymbol{k}$-space is divided into N-patches in fRG.

In Fig. 2.6, the flow of the present fRG+cRPA method is described.

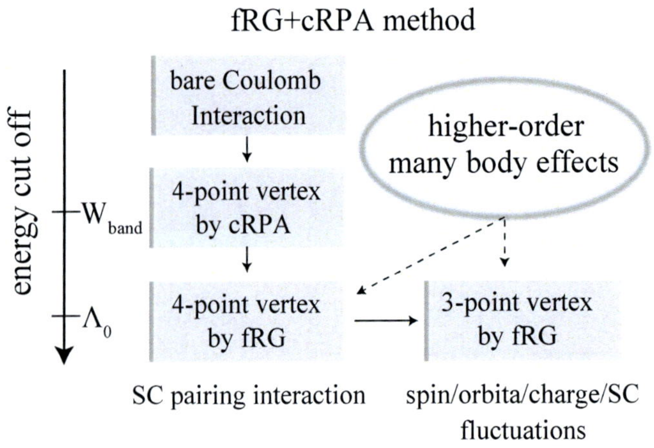

Fig. 2.6 The flow of the fRG+cRPA method

cRPA Method

In high-energy region ($|\xi_k| > \Lambda_0$), we perform the constrained RPA (cRPA). First, d-electrons' Green function is defined as

$$\hat{G}(k) = \frac{1}{i\epsilon_n - \hat{H}_0} = \begin{pmatrix} i\epsilon_n - \xi_k^{11} & -\xi_k^{12} \\ -\xi_k^{21} & i\epsilon_n - \xi_k^{11} \end{pmatrix}^{-1} \quad (2.17)$$

where $k = (\boldsymbol{k}, \epsilon_n) = (\boldsymbol{k}, (2n+1)\pi T)$. $\hat{G}(k)$ is independent of the spin index. Therefore, $\hat{G}(k)$ is 2×2 matrix in the orbital space. By using the Green function, the irreducible susceptibility is calculated by

$$\hat{\chi}^0_{\mathrm{cRPA}}(q) = -T \sum_k \hat{G}(k+q)\theta(|\xi_{\boldsymbol{k}+\boldsymbol{q}}| - \Lambda_0) \cdot \hat{G}(k)\theta(|\xi_{\boldsymbol{k}}| - \Lambda_0), \quad (2.18)$$

where $q = (\boldsymbol{q}, \omega_m) = (\boldsymbol{q}, 2\pi T m)$ and θ is Heaviside step function defined as $\theta(k) = 1(0)$ for $k > 0 (k < 0)$. $\hat{\chi}^0_{\mathrm{cRPA}}$ is $2^2 \times 2^2$ matrix. Then, the spin (charge)-channel susceptibility within the cRPA is obtained by

$$\hat{\chi}^{s(c)}_{\mathrm{cRPA}}(q) = \frac{\hat{\chi}^0_{\mathrm{cRPA}}(q)}{\hat{1} - \hat{U}^{s(c)}\hat{\chi}^0_{\mathrm{cRPA}}(q)}. \quad (2.19)$$

% Then, the four-point vertex function within the cRPA is obtained by

$$\hat{\Gamma}^{s(c)}_{\mathrm{cRPA}}(k+q, k; k'+q, k') = \hat{U}^{s(c)} + \hat{U}^{s(c)}\hat{\chi}^{s(c)}_{\mathrm{cRPA}}(k-k')\hat{U}^{s(c)} - A.C., \quad (2.20)$$

where A.C. denotes anti-commutation of the second term for orbital, spin, and k index. The detailed definition of $\Gamma(k+q, k; k'+q, k')$ is given in Fig. 2.7b. Here,

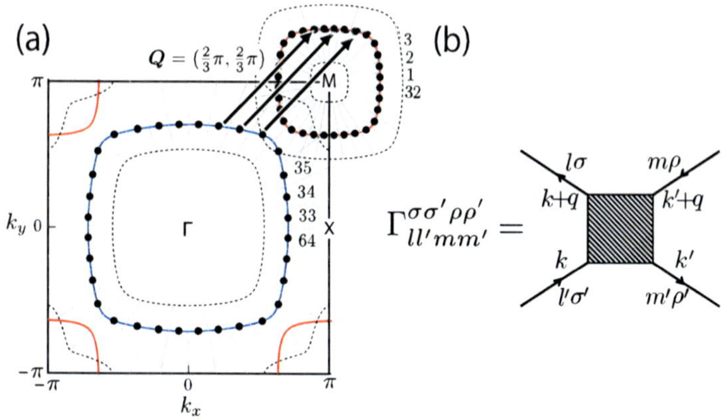

Fig. 2.7 **a** The region of each patches on the Fermi surface. **b** Definition of four-point vertex function in the present study

$\hat{U}^c$ and $\hat{U}^s$ in Eqs. (2.10)–(2.12) are used as the initial values for the four-point vertex. Therefore, the four-point vertex function has $\boldsymbol{k}$-dependence within the cRPA, while it does not have in the initial Hamiltonian. Note that we neglect the Cooper channel scattering in the cRPA since it is negligibly small in the higher energy region, while it gives significant contribution in lower energy region.

fRG Method

Then, we perform the fRG method to obtain fully renormalized four-point vertex function $\hat{\Gamma}_{\rm RG}$ including low-energy scatterings. We use the $\hat{\Gamma}^{s(c)}_{\rm cRPA}$ as a initial value for fRG. The four-point vertex function is automatically obtained by solving the differential equation. The one-loop RG equation in the band basis is given by

$$\frac{d}{d\Lambda}\Gamma_{\rm RG}(k_1, k_2; k_3, k_4) = -T\sum_{k,k'}\frac{d}{d\Lambda}\left[\tilde{G}(k)\,\tilde{G}(k')\right]\Big[\Gamma_{\rm RG}(k_1, k_2; k, k')\Gamma_{\rm RG}(k, k'; k_3, k_4)$$
$$-\Gamma_{\rm RG}(k_1, k_3; k, k')\Gamma_{\rm RG}(k, k'; k_2, k_4) - \frac{1}{2}\Gamma_{\rm RG}(k_1, k; k', k_4)\Gamma_{\rm RG}(k, k_2; k_3, k')\Big], \quad (2.21)$$

where the compact notation k denotes $k = (\boldsymbol{k}, \epsilon_n, u, \sigma)$ and u is band index. The band and orbital representation is connected by the unitary transformation $u_{lu}(\boldsymbol{k}) = \langle l, \boldsymbol{k}|u, \boldsymbol{k}\rangle$. The first and second terms in the r.h.s. of Eq. (2.21) are originating from Peierls-ch scattering, while the third one corresponds to Cooper-ch scatterings. $\tilde{G}(k)$ is defined in the band basis as

$$\tilde{G}(k) \equiv G(k)\theta(|\xi^u_{\boldsymbol{k}}| - \Lambda). \quad (2.22)$$

Then, Eq. (2.21) is rewritten by using on-shell Green function $G_{shell}(\Lambda - d\Lambda \leq |\xi_{\boldsymbol{k}}| < \Lambda)$ and high-energy one $G_{>}(\Lambda \leq |\xi_{\boldsymbol{k}}| < \Lambda_0)$. For instance, the first term of the r.h.s. of Eq. (2.21) is given by

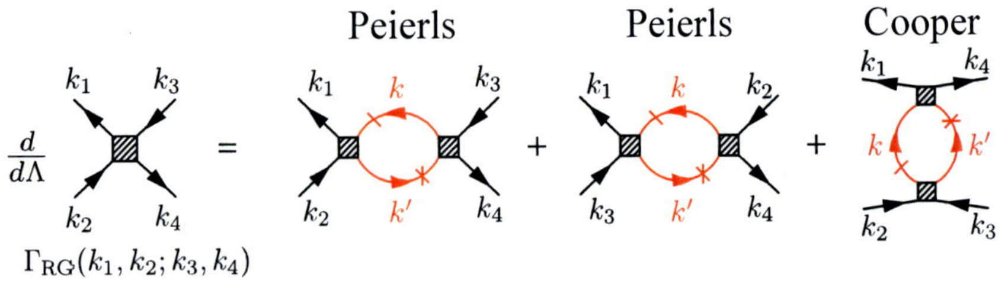

Fig. 2.8 One-loop RG differential equation for four-point vertex function. The crossed red line represents electron propagator G with energy cutoff Λ, while the slashed line denotes on-shell one

$$
\begin{aligned}
&-T\sum_{k,k'}\frac{d}{d\Lambda}\left[\tilde{G}(k)\tilde{G}(k')\right]\left[\Gamma_{\rm RG}(k_1,k_2;k,k')\Gamma_{\rm RG}(k,k';k_3,k_4)\right] \\
&\quad = \int_{|\xi_k|=\Lambda}\frac{dk}{v_k}\Gamma_{\rm RG}(k_1,k_2;k,k-k_1+k_2)\frac{f(\xi_{k-k_1+k_2})-f(\xi_k)}{\xi_{k-k_1+k_2}-\xi_k} \\
&\quad \times\theta(|\xi_{k-k_1+k_2}|-\Lambda)\Gamma_{\rm RG}(k,k-k_1+k_2,k_3,k_4) \\
&\quad + \int_{|\xi_{k'}|=\Lambda}\frac{dk'}{v_{k'}}\Gamma_{\rm RG}(k_1,k_2;k_1-k_2+k',k')\frac{f(\xi_{k'})-f(\xi_{k_1-k_2+k'})}{\xi_{k'}-\xi_{k_1-k_2+k'}} \\
&\quad \times\theta(|\xi_{k_1-k_2+k'}|-\Lambda)\Gamma_{\rm RG}(k_1-k_2+k',k';k_3,k_4).
\end{aligned}
\tag{2.23}
$$

Now, we obtain fully renormalized four-point vertex function. The diagrammatic expression of the RG equation is shown in Fig. 2.8. Here, we neglect the effect of self-energy [19–21]. The energy scale will change if we consider the self-energy based on the Fermi liquid theory. We comment that the present fRG method adopt high-energy Green function in contrast to Wilson's RG method. Both methods will give the same result when the cutoff energy is sufficiently small. It is verified that the present result does not change if we use the low-energy Green function.

Under the SU(2) symmetry, $\hat{\Gamma}_{\rm RG}$ in the orbital basis is uniquely decomposed into spin and charge channels as

$$
\begin{aligned}
\hat{\Gamma}_{\rm RG}(k+q,k;k'+q,k') &= \frac{1}{2}\Gamma^s_{\rm RG}(k+q,k;k'+q,k')\boldsymbol{\sigma}_{\sigma\sigma'}\cdot\boldsymbol{\sigma}_{\rho'\rho} \\
&+ \frac{1}{2}\Gamma^c_{\rm RG}(k+q,k;k'+q,k')\delta_{\sigma\sigma'}\delta_{\rho'\rho}.
\end{aligned}
\tag{2.24}
$$

As a result, parquet-type scattering processes are automatically generated as shown in Fig. 2.9. In principle, the present fRG is based on Ward-Takahashi identity, which gives the relation between N- and N+2-point vertex functions while we neglect higher than six-point vertex function and two-loop diagram.

Merit of fRG+cRPA

Compared with the previous other fRG study, the present fRG+cRPA method shows high accuracy by using the cRPA. Moreover, the algorithm for identifying the momen-

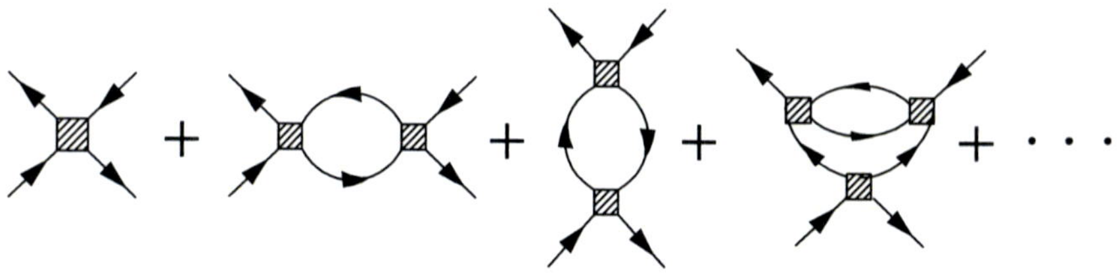

Fig. 2.9 Parquet-type diagrams considered in the one-loop RG equation

tum conservation condition $k_1 + k_4 \simeq k_2 + k_3$ is improved in the present study as explained in Appendix 1. One of the great merits of the fRG is dealing with both Cooper- and Peierls-ch scatterings in the same footing. In fact, perfect cancellation occurs between Cooper- and Peierls-ch scatterings in 1D system. Therefore, both channels are important in the low-dimensional system. The great merits of the present fRG+cRPA method are summarized as follows:

- Higher order many body effects are automatically considered by the fRG.
- High accuracy is realized by the combination of fRG and cRPA.
- Both Cooper and Peierls-ch scatterings are considered in fRG.

2.2.3 *Three-Point Vertex Function*

Peierls-ch Susceptibility

Here, we calculate the d-electrons' susceptibility within the fRG+cRPA method. The definition of charge- (spin-) channel (ch) susceptibility in the orbital basis is given by

$$\chi^{c(s)}_{ll'mm'}(q) = \int_0^\beta d\tau \frac{1}{2}\left\langle A^{c(s)}_{ll'}(\boldsymbol{q},\tau)A^{c(s)}_{m'm}(-\boldsymbol{q},0)\right\rangle e^{i\omega_l\tau}, \tag{2.25}$$

$$A^{c(s)}_{ll'}(\boldsymbol{q}) \equiv \sum_{\boldsymbol{k}}(d^{\dagger}_{\boldsymbol{k}l'\uparrow}d_{\boldsymbol{k}+\boldsymbol{q}l\uparrow} + (-)d^{\dagger}_{\boldsymbol{k}l'\downarrow}d_{\boldsymbol{k}+\boldsymbol{q}l\downarrow}). \tag{2.26}$$

In the framework of fRG theory, the susceptibilities are calculated by solving the differential equation (Fig. 2.10a),

$$\frac{d}{d\Lambda}\chi^{c(s)}_{\mathrm{RG}}(q) = T\sum_k \frac{d}{d\Lambda}\left[\tilde{G}(k)\tilde{G}(k+q)\right]R^{c(s)}(q;k,k+q)R^{c(s)}(-q;k+q,k), \tag{2.27}$$

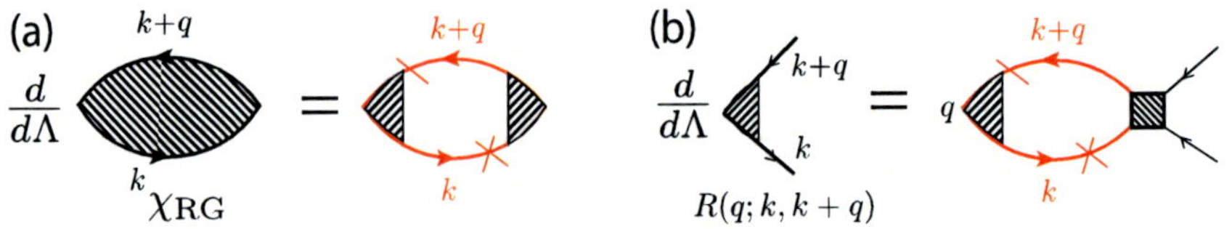

Fig. 2.10 **a** RG equation for susceptibility. The shaded triangles are three-point vertex function. **b** RG equation for three-point vertex function

where $R^{c(s)}$ is the three-point vertex diagrammatically expressed in Fig. 2.10b. The three-point vertex function $R^{c(s)}$ is derived from

$$\frac{d}{d\Lambda}R^{c(s)}(q;k,k') = -T\sum_{k''}\frac{d}{d\Lambda}\left[\tilde{G}(k'')\tilde{G}(k''+q)\right] \times R^{c(s)}(q;k'',k''+q)\Gamma^{c(s)}_{\mathrm{RG}}(k,k''+q,k',k''). \quad (2.28)$$

The initial value for the susceptibility and three-point vertex function is calculated by the cRPA in the same way for the four-point vertex. Then, we obtain charge-(spin-)ch susceptibility in the orbital basis given by

$$\hat{\chi}^{c(s)}_{\mathrm{RG}}(q) = \hat{\Phi}^{c(s)}(q)[\hat{1} - \hat{U}^{c(s)}\hat{\Phi}^{c(s)}(q)]^{-1}$$
$$\hat{\Phi}^{c(s)}(q) \equiv -T\sum_{n}\hat{G}(k+q)\hat{G}(k)\hat{\Lambda}^{c(s)}(k+q,k), \quad (2.29)$$

where $\hat{\Lambda}$ is VC for irreducible susceptibility due to many body effects beyond RPA. Hereafter, we call $\hat{\Lambda}$ χ-VC. The great merit of fRG method is that χ-VC is automatically calculated just by solving the renormalization equation.

Cooper-ch Susceptibility

Now, we introduce the Cooper-ch susceptibilities in the band basis defined by

$$\chi^{\mathrm{SC}}_{t(s)} = \frac{1}{2}\int_0^\beta d\tau \left\langle B^\dagger_{t(s)}(\tau)B_{t(s)}(0)\right\rangle, \quad (2.30)$$
$$B_{t(s)} \equiv \sum_{\boldsymbol{k}}\Delta_{t(s)}(\boldsymbol{k})d_{\boldsymbol{k}\uparrow}d_{-\boldsymbol{k}\uparrow(\downarrow)},$$

where $\Delta_{t(s)}(\boldsymbol{k})$ is TSC (SSC) gap function on the Fermi surface. The $\boldsymbol{k}$-dependence of the gap function is uniquely determined so as to maximize the Cooper-ch susceptibilities.

- Peierls- and Cooper-ch susceptibilities are derived from the three-point vertex function.
- Higher order many body effects given by χ-VC are automatically generated.

2.3 Results

In this section, we show numerical results obtained by fRG+cRPA. We put the parameter $T = 5 \times 10^{-4}$ and low-energy cutoff $\Lambda_l = 0.01T$ (i.e., $l = \ln(\Lambda_0/0.01T)$) in the energy unit $|t| = 1$.

2.3.1 *Emergence of Triplet Superconductivity*

Spin/Orbital Susceptibility

First, we show the obtained result of spin- and charge-ch susceptibilities at $(U, J/U) = (3.10, 0.08)$ in Fig. 2.11a, b, respectively. In Fig. 2.11a, obtained total spin susceptibility is plotted, which is defined as

$$\chi^s(\boldsymbol{q}) = \sum_{l,m} \chi^s_{llmm}(\boldsymbol{q}). \tag{2.31}$$

In addition, we obtain large quadrupole susceptibility as shown in Fig. 2.11b, which is given by

$$\chi^c_{x^2-y^2}(\boldsymbol{q}) = \sum_{l,m} (-1)^{l+m} \chi^c_{llmm}(\boldsymbol{q}). \tag{2.32}$$

The quadrupole susceptibility diverges when orbital polarization $n_{xz} - n_{yz}$ emerges. Both $\chi^s(\boldsymbol{q})$ and $\chi^c_{x^2-y^2}(\boldsymbol{q})$ have peak at the nesting vector $\boldsymbol{Q} = (2\pi/3, 2\pi/3)$. Large spin fluctuation at $\boldsymbol{Q} = (2\pi/3, 2\pi/3)$ is observed by neutron scattering [22] in Sr_2RuO_4. Therefore, we obtain large orbital fluctuations as well as AFM fluctuations,

$$\chi^s(\boldsymbol{Q}) \approx \chi^c_{x^2-y^2}(\boldsymbol{Q}) \quad \text{(fRG)}, \tag{2.33}$$

while $\chi^c(\boldsymbol{Q}) \ll \chi^s(\boldsymbol{Q})$ is realized within RPA. Thus, we conclude that the large orbital fluctuations are induced by many body effects due to the χ-VC considered in the fRG.

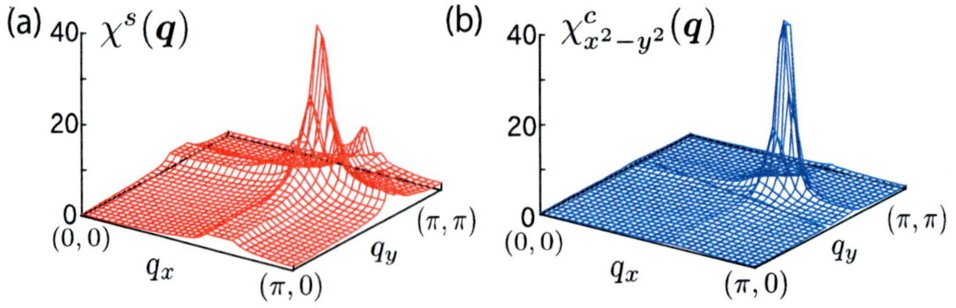

Fig. 2.11 **a** Momentum dependence of total spin susceptibility $\chi^s(\boldsymbol{q})$ and **b** that of quadrupole susceptibility $\chi^c_{x^2-y^2}(\boldsymbol{q})$, which show strong peak at the nesting vector $\boldsymbol{q} \approx (2\pi/3, 2\pi/3)$. Reprinted with permission from Ref. [11]. Copyright ©2016 by the American Physics Society

Phase Diagram

We show the obtained phase diagram in Fig. 2.12. The boundary of the orbital- and magnetic-ordered phase is written by the broken lines, while the solid lines represent that of spin triplet and singlet SC phase. The relation $\chi^s(\boldsymbol{Q}) = \chi^c_{x^2-y^2}(\boldsymbol{Q})$ holds on the dotted line. The boundary line is determined by the divergence point of Peierls- and Cooper-ch susceptibilities.

We reveal that spin triplet and singlet SC phase appear near the orbital- and magnetic-ordered phase in wide parameter region, respectively. In particular, it is predicted that TSC phase appears as an intertwined order between orbital-ordered

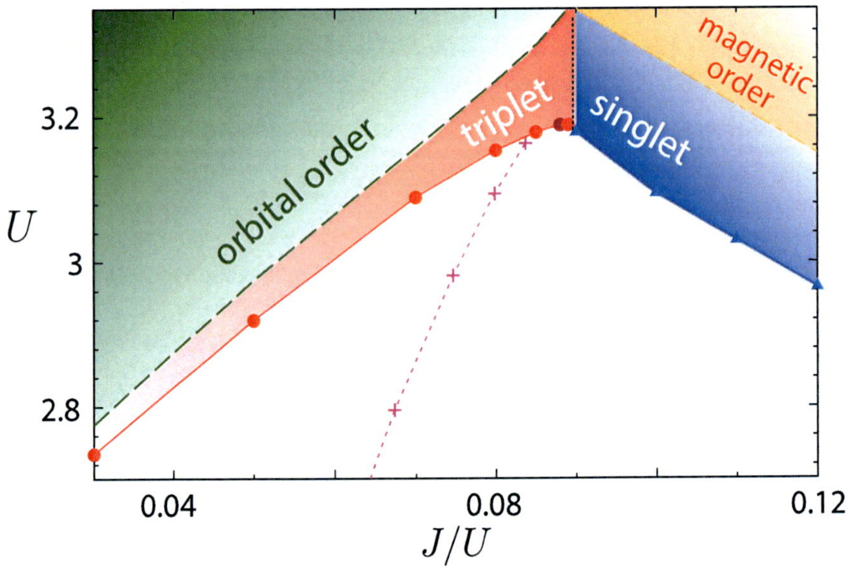

Fig. 2.12 Obtained phase diagram by fRG+cRPA method

and AFM phases. Obtained TSC gap $\Delta_t(k)$ belongs to the E_u representation and it is written as

$$\Delta_{t,x}(\boldsymbol{k}),\ \Delta_{t,y}(\boldsymbol{k}) \propto \sin 3k_x,\ \sin 3k_y \quad (\in E_{1u}),$$
$$\Delta_s(\boldsymbol{k}) \propto 1 \quad (\in A_{1g}). \tag{2.34}$$

Spin singlet SC gap $\Delta_s(k)$ is in the A_{1g} symmetry. We found that the strong orbital fluctuations develop at $J/U \lesssim 0.1$. This value is reasonable compared with $J/U = 9.5 \times 10^{-2}$ reported in FeSe by the first-principles study.

- Orbital fluctuations are induced by χ-VC in fRG method.
- Triplet SC phase appears near the boundary of orbital-ordered and AFM phases.

2.3.2 SC Pairing Interaction

Here, we analyze the SC pairing interaction based on the linearized gap equation [23]. The linearized SC gap equation on the Fermi surface is written by

$$\lambda_{t(s)}\Delta_{t(s)}(\boldsymbol{k}) = -\int \frac{d\boldsymbol{k}'}{v_{\boldsymbol{k}'}} V^{\omega_c}_{t(s)}(\boldsymbol{k},\boldsymbol{k}')\Delta_{t(s)}(\boldsymbol{k}') \ln\frac{1.13\omega_c}{T}. \tag{2.35}$$

Here, the momenta $\boldsymbol{k}, \boldsymbol{k}'$ is on the Fermi surface. $v_{\boldsymbol{k}}(= d\xi_{\boldsymbol{k}}/dk)$ is Fermi velocity. $\lambda_{t(s)}$ gives eigenvalue for triplet (singlet) SC. $V^{\omega_c}_{t(s)}$ is effective SC pairing interaction by fRG+cRPA. When we consider only even-frequency gap function, TSC (SSC) gap has odd (even) parity and

$$\Delta_t(\boldsymbol{k}) = -\Delta_t(-\boldsymbol{k}), \quad \Delta_s(\boldsymbol{k}) = \Delta_s(-\boldsymbol{k}). \tag{2.36}$$

In the derivation of Eq. (2.35), we used the general relation

$$\int_{-\omega_c}^{\omega_c} d\epsilon_{\boldsymbol{k}'} \frac{1}{2\epsilon_{\boldsymbol{k}'}} \mathrm{th}(\epsilon_{\boldsymbol{k}'}/2T) = \ln(1.13\omega_c/T). \tag{2.37}$$

By solving the SC gap equation in Eq. (2.35), we obtain $\lambda_t = 0.47$ and $\lambda_s = 0.26$. In addition, the triplet (singlet) gap function belongs to E_{1u} (A_{1g}) representation, which is consistent with the SC fluctuations. Therefore, solving the SC gap equation is essentially equivalent to calculating the SC fluctuations based on the fRG.

Pairing Interaction with U-VC

The SC pairing interaction is directly calculated from four-point vertex $\Gamma_{\rm RG}$ as

$$V_{t,\rm RG}^{\omega_c}(\boldsymbol{k},\boldsymbol{k}') = -\frac{1}{4}\Gamma_{\rm RG}^{s}(\boldsymbol{k},\boldsymbol{k}';-\boldsymbol{k}',-\boldsymbol{k}) - \frac{1}{4}\Gamma_{\rm RG}^{c}(\boldsymbol{k},\boldsymbol{k}';-\boldsymbol{k}',-\boldsymbol{k}), \quad (2.38)$$

$$V_{s,\rm RG}^{\omega_c}(\boldsymbol{k},\boldsymbol{k}') = \frac{3}{4}\Gamma_{\rm RG}^{s}(\boldsymbol{k},\boldsymbol{k}';-\boldsymbol{k}',-\boldsymbol{k}) - \frac{1}{4}\Gamma_{\rm RG}^{c}(\boldsymbol{k},\boldsymbol{k}';-\boldsymbol{k}',-\boldsymbol{k}). \quad (2.39)$$

The four-point vertex has low-energy cutoff $\Lambda_l = \omega_c$. In the present study, we set $\omega_c = 12T = 6\times 10^{-3}$ so as to satisfy $\omega_c > T$. This condition is reasonable since Peierls susceptibilities saturate at the energy scale of T, while Cooper-ch ones drastically develop in $\Lambda_l < \omega_c$.

The pairing interaction includes higher order many body effects beyond Migdal-Eliashberg approximation. The diagrammatic expression of the pairing interaction is given in Fig. 2.13. The filled triangles show the vertex corrections (VC) for SC pairing interaction. We call this U-VC, which is neglected in Migdal-Eliashberg (ME) scheme. However, we reveal that U-VC becomes important in low-energy region at $\omega_c \ll T_{\rm c}$. Typical diagrams of U-VC are represented in Fig. 2.13b. Aslamazov-Larkin (AL) and Maki-Thompson (MT) terms are included.

Pairing Interaction without U-VC

In order to show the importance of U-VC clearly, we calculate SC interaction without U-VC. For this purpose, $\hat{\chi}_{\rm RG}^{c(s)}$ with cutoff $\Lambda_l = \omega_c$ is used to obtain the pairing interaction, which is given by

$$\hat{V}_{s,\chi}(\boldsymbol{k},\boldsymbol{k}') \equiv \frac{3}{4}\hat{\Gamma}_{\chi}^{s}(\boldsymbol{k},\boldsymbol{k}') - \frac{1}{4}\hat{\Gamma}_{\chi}^{c}(\boldsymbol{k},\boldsymbol{k}') \quad (2.40)$$

$$\hat{V}_{t,\chi}(\boldsymbol{k},\boldsymbol{k}') \equiv -\frac{1}{4}\hat{\Gamma}_{\chi}^{s}(\boldsymbol{k},\boldsymbol{k}') - \frac{1}{4}\hat{\Gamma}_{\chi}^{c}(\boldsymbol{k},\boldsymbol{k}') - \frac{1}{2}\hat{U}^{s}. \quad (2.41)$$

The four-point vertex function Γ_χ is derived from

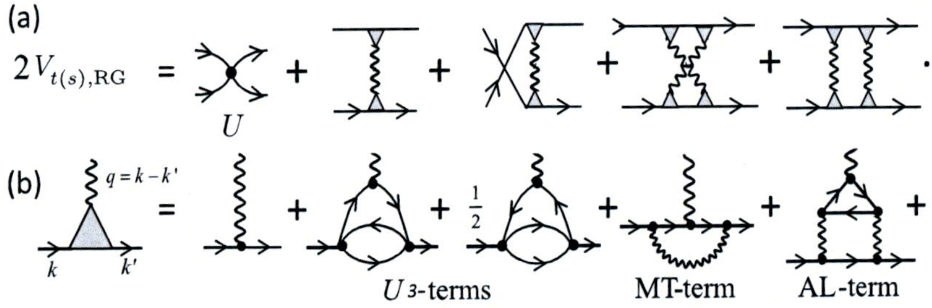

Fig. 2.13 **a** SC pairing interaction given by fRG method. **b** Typical diagrams of the U-VC. Double counting is carefully avoided

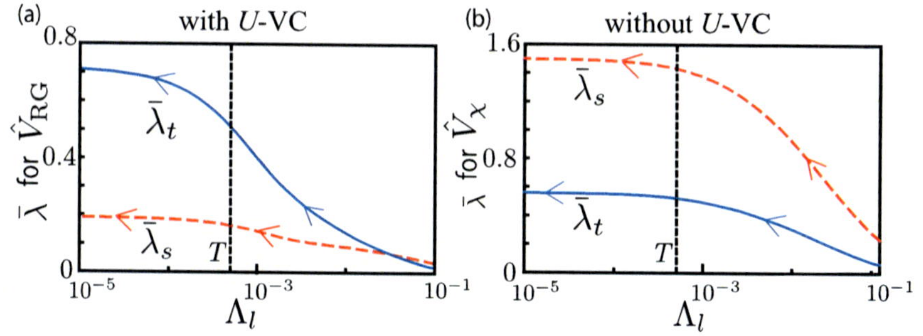

Fig. 2.14 **a** Effective SC pairing interaction $\bar{\lambda}$ calculated by $V_{t(s),\mathrm{RG}}$ and **b** $V_{t(s),\chi}$. Reprinted with permission from Ref. [11].

$$\Gamma^{s}_{\chi,ll'mm'}(\boldsymbol{k},\boldsymbol{k}') = \left(\hat{U}^s + \hat{U}^s\hat{\chi}^s_{\mathrm{RG}}(\boldsymbol{k}-\boldsymbol{k}')\hat{U}^s\right)_{ll'mm'} - \left(\frac{1}{2}\hat{U}^c\hat{\chi}^c_{\mathrm{RG}}(\boldsymbol{k}+\boldsymbol{k}')\hat{U}^c - \frac{1}{2}\hat{U}^s\hat{\chi}^s_{\mathrm{RG}}(\boldsymbol{k}+\boldsymbol{k}')\hat{U}^s\right)_{lml'm'}, \quad (2.42)$$

$$\Gamma^{c}_{\chi,ll'mm'}(\boldsymbol{k},\boldsymbol{k}') = \left(\hat{U}^c + \hat{U}^c\hat{\chi}^c_{\mathrm{RG}}(\boldsymbol{k}-\boldsymbol{k}')\hat{U}^c\right)_{ll'mm'} - \left(\frac{1}{2}\hat{U}^c\hat{\chi}^c_{\mathrm{RG}}(\boldsymbol{k}+\boldsymbol{k}')\hat{U}^c + \frac{3}{2}\hat{U}^s\hat{\chi}^s_{\mathrm{RG}}(\boldsymbol{k}+\boldsymbol{k}')\hat{U}^s\right)_{lml'm'}. \quad (2.43)$$

In this case, U-VC is neglected in the SC pairing interaction, while χ-VC is included.

Comparison: with U-VC Versus Without U-VC
In Fig. 2.14, we show the effective pairing interaction $\bar{\lambda}_{t(s)}$ defined by

$$\bar{\lambda}_{t(s)} = \frac{\displaystyle\int \frac{d\boldsymbol{k}}{v_{\boldsymbol{k}}} \int \frac{d\boldsymbol{k}'}{v_{\boldsymbol{k}'}} V^{\omega_c}_{t(s)}(\boldsymbol{k},\boldsymbol{k}')\Delta_{t(s)}(\boldsymbol{k}')\Delta_{t(s)}(\boldsymbol{k})}{\displaystyle\int \frac{d\boldsymbol{k}}{v_{\boldsymbol{k}}}\Delta_{t(s)}(\boldsymbol{k})\Delta_{t(s)}(\boldsymbol{k})}, \quad (2.44)$$

where the $\boldsymbol{k}$ is the momenta on the Fermi surface. $\bar{\lambda}_{t(s)}$ is related to the T_c as $T_{c,t(s)} \simeq 1.13\omega_c \exp(-1/\bar{\lambda}_{t(s)})$. Note that $\bar{\lambda}_{t(s)}$ almost saturates when Λ_l is lower than T. As a result, we obtain the following results:

$$\begin{aligned} \bar{\lambda}_t &\sim 3\bar{\lambda}_s \ \text{(with } U\text{-VC)} \\ \bar{\lambda}_s &\gg \bar{\lambda}_s \ \text{(without } U\text{-VC).} \end{aligned} \quad (2.45)$$

Therefore, TSC is mediated by the important role of U-VC. Note that $\chi^s(\boldsymbol{Q}) \sim \chi^c_{x^2-y^2}(\boldsymbol{Q})$ even in the case of $\bar{\lambda}_t \sim 3\bar{\lambda}_s$. Thus, moderate orbital fluctuation is enough to realize the TSC phase. We summarize the present results in Fig. 2.15.

	U-VC	χ-VC	SC phase
V_{RG}	○	○	**triplet**
V_{χ}	×	○	**singlet**
$V_{RPA+U\text{-}VC}$	○	×	**singlet**
V_{RPA+ME}	×	×	**singlet**

Fig. 2.15 Summary of the present results. TSC appears in the presence of both U-VC and χ-VC

- SC pairing interaction by fRG includes both χ-VC and U-VC.
- SC pairing interaction without U-VC is obtained by $\hat{\chi}_{\mathrm{RG}}^{c(s)}$.
- Triplet SC state is stabilized by the significant roles of χ-VC and U-VC.

2.3.3 Significant Roles of U-VC

Momentum Dependence of U-VC

To understand the origin of U-VC, we focus on the momentum dependence of U-VC. Figure 2.16a–d shows the inter-band component of the spin- and charge-channel SC pairing interactions. Note that intra-band term is much smaller than inter-band one. Here, i_α and i_β are patch indexes on Fermi surface α and β, respectively. The peaks in solid ellipsoid area are originating from the nesting vector at $\boldsymbol{Q}$. We find the relation that

$$\Gamma_{\chi}^{s}(\boldsymbol{k}, \boldsymbol{k}') \gg \Gamma_{\chi}^{c}(\boldsymbol{k}, \boldsymbol{k}') \ \ (\text{without} U\text{-VC}). \tag{2.46}$$

Therefore, the repulsion due to χ^s is much larger than attraction due to the χ^c. For this reason, TSC does not appear if we neglect the U-VC.

On the other hand, the situation will drastically change by considering U-VC. Figure 2.16c, d shows $\Gamma_{\mathrm{RG}}^{s}(\boldsymbol{k}, \boldsymbol{k}')$ and $\Gamma_{\mathrm{RG}}^{c}(\boldsymbol{k}, \boldsymbol{k}')$ in the presence of U-VC. Both spin- and charge-channel interactions show large positive values at $\boldsymbol{Q}$. Moreover, we obtain the relation

$$\Gamma_{\chi}^{s}(\boldsymbol{k}, \boldsymbol{k}') \ll \Gamma_{\chi}^{c}(\boldsymbol{k}, \boldsymbol{k}') \ \ (\text{with } U\text{-VC}). \tag{2.47}$$

Therefore, the spin- (charge-) channel pairing interaction is drastically suppressed (enhanced) by U-VC. As a result, orbital fluctuation-driven TSC phase emerges based on the fRG. Furthermore, the momentum dependence of $\Gamma_{\mathrm{RG}}(\boldsymbol{k}, \boldsymbol{k}')$ and $\Gamma_{\chi}(\boldsymbol{k}, \boldsymbol{k}')$ is very similar. This fact means that single-fluctuation-exchange term is quite important since multi-fluctuation-exchange process gives different momentum dependence. We

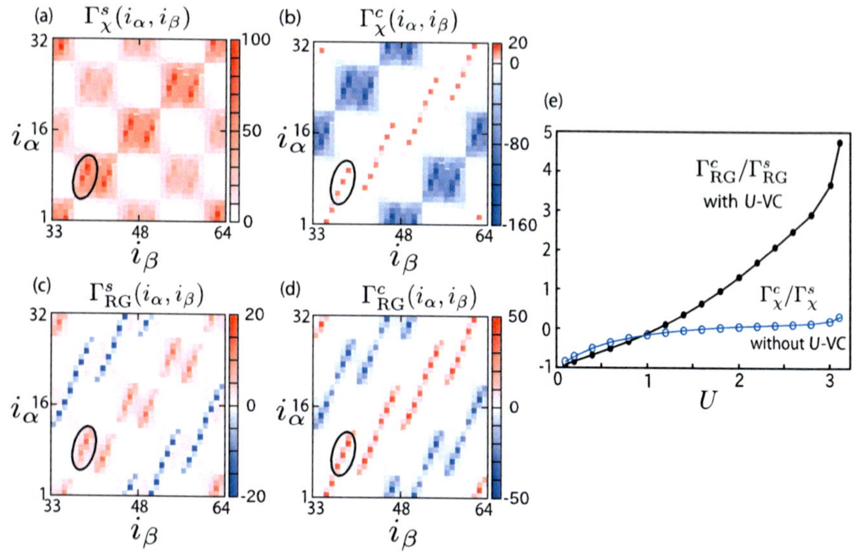

Fig. 2.16 **a**, **b** Inter-band SC pairing interactions without U-VC and **c**, **d** that with U-VC. **e** The obtained ratio of Γ^c/Γ^s as a functions of U. Reprinted with permission from Ref. [11]. Copyright ©2016 by the American Physics Society

comment that large negative value at $(i_\alpha, i_\beta) = (6+16, 37), (8+16, 38), (10+16, 39)$ comes from spin fluctuations at $\boldsymbol{k}+\boldsymbol{k}' \approx \boldsymbol{Q}$ given by third terms of the r.h.s. in Eqs. (2.42) and (2.43).

U-dependence of U-VC
Figure 2.16e shows the ratio of $\Gamma^c_\chi/\Gamma^s_\chi$ and $\Gamma^c_{\rm RG}/\Gamma^s_{\rm RG}$ at $\boldsymbol{Q}$ as a functions of U. In weakly correlated region around, the ratios take -1, which is originating from

$$\Gamma^s_\chi, \Gamma^s_{\rm RG} \approx U \ , \ \Gamma^c_\chi, \Gamma^c_{\rm RG} \approx -U \ \ (U \approx 0). \tag{2.48}$$

Therefore, the bare Coulomb interaction is dominant in weakly coupled region. In contrast, the ratio of $\Gamma_{\rm RG}$ increases around while that of Γ_χ remains small summarized as

$$\Gamma^c_\chi/\Gamma^s_\chi \approx const. \ , \ \Gamma^c_{\rm RG}/\Gamma^s_{\rm RG} \approx \text{large} \ \ (U \gtrsim 2). \tag{2.49}$$

Thus, the four-point vertex $\Gamma^{c(s)}_{\rm RG}$ is enhanced (suppressed) in the strong coupling region. In conclusion, the spin-channel SC pairing interaction is drastically suppressed, while the charge-channel one is enhanced by the U-VC.

- Inter-band SC pairing interaction is important for the present TSC scenario.
- Single-fluctuation-exchange process brings dominant contribution.
- U-VC enhances (suppresses) the pairing interaction due to the orbital (spin) fluctuations.

2.3.4 Perturbation Theory

In this section, we perform diagrammatic analysis to understand the microscopic origin of the phase diagram by fRG+cRPA. In particular, we consider MT- and AL-terms for the U-VC. The MT-term is the first order of the spin- and charge-channel fluctuations given by

$$U^{c,\mathrm{MT}}(q) \propto \left\{\chi^c(q) + 3\chi^s(q)\right\}$$
$$U^{s,\mathrm{MT}}(q) \propto \left\{\chi^c(q) - \chi^s(q)\right\}. \tag{2.50}$$

The AL-term is the second order of the fluctuation given by

$$U^{c,\mathrm{AL}}(q) \propto \sum_p \left\{\chi^c(q+p)\chi^c(q) + 3\chi^s(q+p)\chi^s(q)\right\}$$
$$U^{s,\mathrm{AL}}(q) \propto \sum_p \left\{2\chi^s(q+p)\chi^c(q)\right\}. \tag{2.51}$$

The detailed expression is given in Appendix 1.

Figure 2.17a shows the enhancement factor due to the AL-term and MT-term, which is defined by

$$r^{c(s)} \equiv \left(\frac{U^{c(s)}(\boldsymbol{k},\boldsymbol{k}') + U^{c(s),\mathrm{AL}}(\boldsymbol{k},\boldsymbol{k}') + U^{c(s),\mathrm{MT}}(\boldsymbol{k},\boldsymbol{k}')}{U^{c(s)}(\boldsymbol{k},\boldsymbol{k}')}\right)^2. \tag{2.52}$$

Here, α_S is spin Stoner factor defined by the largest eigenvalue of $\hat{U}^s\hat{\chi}^0(\boldsymbol{q})$ and thus $\chi^s(\boldsymbol{Q}) \propto (1-\alpha_S)^{-1}$. Note that α_S reaches unity when the magnetic transition occurs.

As shown in Fig. 2.17a, r^c gradually increases near the magnetic QCP ($\alpha_S \approx 1$). For this reason, we find that r_c is originating from the AL-terms.

$$r_c \propto \sum_{\boldsymbol{q}} \chi^s(\boldsymbol{q}+\boldsymbol{Q})\chi^s(\boldsymbol{q}) \sim (1-\alpha_S)^{-1} \quad \text{(AL-term)}. \tag{2.53}$$

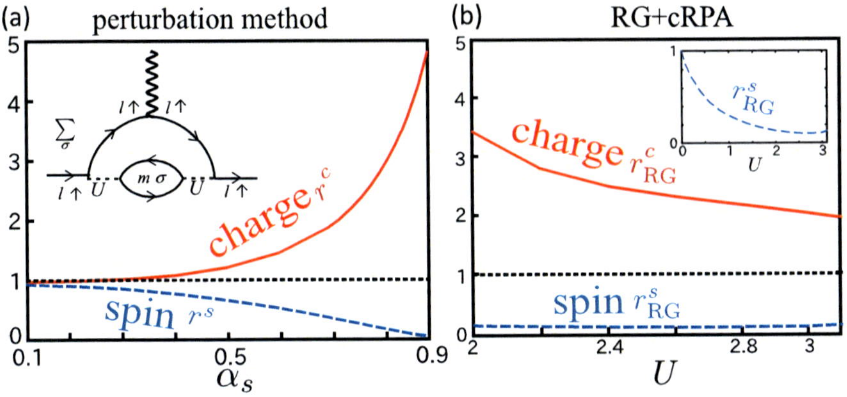

Fig. 2.17 **a** Enhancement factor r for SC pairing interaction due to the AL- and MT-terms. **b** r based on fRG+cRPA. Reprinted with permission from Ref. [11]. Copyright ©2016 by the American Physics Society

In spite of that, suppression of r_s occurs due to the $O(U^3)$-term,

$$r_s \propto (2N_{\rm orb} - 1) \ \ (O(U^3)\text{-term}), \tag{2.54}$$

where $N_{\rm orb}$ is the number of orbitals. Thus, $O(U^3)$-term becomes significant in multi-orbital system. The diagrammatic expression of the $O(U^3)$-term is shown in Fig. 2.17a. We comment that spin-channel AL-term is negligibly small since it is proportional to $\sum_q \chi^s(\boldsymbol{q})\chi^c(\boldsymbol{q}+\boldsymbol{Q})$.

Figure 2.17b shows enhancement factor directly obtained by fRG+cRPA,

$$r^{c(s)}_{\rm RG} \equiv \frac{\Gamma^{c(s)}_{\rm RG}(\boldsymbol{k},\boldsymbol{k}')}{\Gamma^{c(s)}_{\chi}(\boldsymbol{k},\boldsymbol{k}')}. \tag{2.55}$$

Note that $r^{c(s)}_{\rm RG}$ is proportional to $(U\text{-VC})^2$, which is originating from single-fluctuation-exchange term. The obtained result is consistent with that by perturbation study. Thus, the fRG result is understood by the AL- and MT-type vertex corrections.

As a result, we conclude that the SC pairing interaction due to the orbital fluctuations is drastically enhanced by AL-terms while the spin fluctuation is suppressed due to the multi-orbital nature.

- Charge-channel pairing interaction is strongly enhanced by AL-type U-VC.
- Spin-channel pairing interaction is suppressed by $O(U^3)$-term.

2.4 Summary

In this section, we study two-orbital Hubbard model by performing the fRG+cRPA method to understand fundamental properties of many body effects beyond RPA+ Migdal-Eliashberg theory. As a result, we reveal that orbital-fluctuation-mediated spin triplet SC phase emerges due to the significant roles of the vertex correction, which we call χ-VC and U-VC. Especially, χ-VC enhances the orbital fluctuations and orbital-ordered phase appears in the realistic parameter region, while only spin-ordered phase emerges within RPA study. Thus, we conclude that χ-VC beyond RPA is important in multi-orbital electron systems. In addition, to go beyond the Migdal-Eliashberg scheme, we analyze SC pairing interaction with VC, which is called U-VC. Owning to the U-VC, orbital-fluctuation-mediated SC interaction is magnified in the strong coupling regime. In particular, AL-type U-VC enhances the orbital-fluctuation-mediated SC interaction. On the other hand, spin-fluctuation-mediated SC interaction is significantly suppressed by $O(U^3)$-term. This suppression does not contradict development of spin fluctuations in multi-orbital system since U-VC is important only at low energy, while spin fluctuations develop from high energy.

We also revealed that the significant contribution comes from the single-fluctuation-exchange term. Therefore, significance of U-VC is clearly verified. As a result, SC phase induced by orbital fluctuations will appear in various multi-orbital SCES, such as in Fe-, Ru-based, and organic superconductor.

Also, the present study predicts that attractive pairing interaction mediated by electron-phonon coupling will also be enhanced by charge-channel U-VC. For instance, T_c of single-layer FeSe can be increased.

The main results of the present study are summarized as follows:

- Based on the fRG+cRPA study, we revealed that many body effects beyond Migdal-Eliashberg theory play important roles in multi-orbital superconductor.
- χ-VC enhances the orbital fluctuations.
- AL-term enhances the charge-channel attractive pairing interaction.
- Orbital-fluctuation-mediated TSC phase appears due to the U-VC when $\chi^s \approx \chi^c$.

Appendix 1

Momentum Conservation
Here, we discuss momentum conservation in the fRG study. By introducing the concept of the patch (p), the four-point vertex function Γ_{p_1,p_2,p_3,p_4} is required to satisfy the momentum conservation given by

$$p_1 + p_4 \simeq p_2 + p_3. \tag{2.56}$$

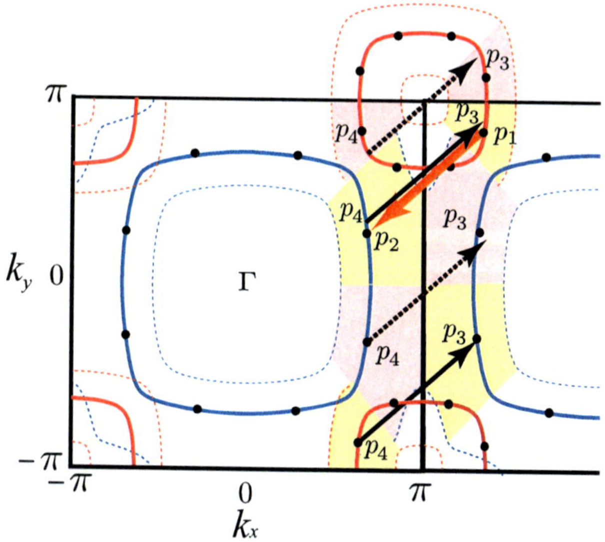

Fig. 2.18 The momentum conservation rule in the present fRG method. Both pink- and yellow-colored regions are considered as the place which p_3 and p_4 belong to

Compared with the common fRG method, we adopt wider condition for the momentum conservation. In the previous study, p_3 and p_4 belong to only yellow-colored area if we put $p_1(p_2)$ at the starting (ending) point of the red arrow given in Fig. 2.18. Therefore, only inter-band scatterings are considered. On the other hand, in the present fRG, we also consider the pink-colored area connected by the black dotted arrows. This contribution was neglected in the previous numerical studies. Thus, low-energy scattering is calculated with higher accuracy than previous works.

Expression of U-VC

Here, we explain analytical expression of the AL- and MT-terms in U-VC. First, we introduce the four-point vertex function in the orbital basis as

$$\hat{I}^{c(s)}(q) = \hat{U}^{c(s)}(\hat{\chi}^{c(s)}_{\rm RPA}(q) + \{\hat{U}^{c(s)}\}^{-1})\hat{U}^{c(s)}. \tag{2.57}$$

By using the four-point vertex, dressed Coulomb interaction due to the MT-type U-VC is given by

$$U^{c,\rm MT}_{ll'mm'}(k,k') = \frac{T}{2}\sum_{q,a\sim d} U^c_{lmbc}\left\{I^c_{amdm'}(q) + 3I^s_{amdm'}(q)\right\} \times G_{ab}(k+q)G_{cd}(k'+q), \tag{2.58}$$

$$U^{s,\rm MT}_{ll'mm'}(k,k') = \frac{T}{2}\sum_{q,a\sim d} U^s_{lmbc}\left\{I^c_{amdm'}(q) - I^s_{amdm'}(q)\right\} \times G_{ab}(k+q)G_{cd}(k'+q), \tag{2.59}$$

where $a \sim h, l, l', m, m'$ are orbital indices. The diagrammatic expression for the MT-term is shown in the second term of the r.h.s. in Fig. 2.13b. Also, the AL-type U-VC is defined by

$$
\begin{aligned}
U^{c,\mathrm{AL}}_{ll'mm'}(k,k') &= \frac{T}{2}\sum_{q,a\sim h} U^{c}_{lmaf}\left\{\Lambda_{abcdef}(k'',q) + \Lambda_{fcbeda}(k'',-q-k'')\right\} \\
&\times \left\{I^{c}_{bcmg}(q+k'')I^{c}_{m'hed}(q) + 3I^{s}_{bcmg}(q+k'')I^{s}_{m'hed}(q)\right\} \\
&\times G_{gh}(k'-q), \\
U^{s,\mathrm{AL}}_{ll'mm'}(k,k') &= \frac{T}{2}\sum_{q,a\sim h} U^{s}_{lmaf}\left\{\Lambda_{abcdef}(k'',q) + \Lambda_{fcbeda}(k'',-q-k'')\right\} \\
&\times \left\{I^{s}_{bcmg}(q+k'')I^{c}_{m'hed}(q) + I^{c}_{bcmg}(q+k''))I^{s}_{m'hed}(q)\right\} \\
&\times G_{gh}(k'-q) + \delta U^{s,\mathrm{AL}}_{ll'mm'}(k,k'), \\
\delta U^{s,\mathrm{AL}}_{ll'mm'}(k,k') &= \frac{T}{2}\sum_{q,a\sim h} U^{s}_{ll'mm'}\left\{\Lambda_{abcdef}(k'',q) - \Lambda_{fcbeda}(k'',-q-k'')\right\} \\
&\times \left\{2I^{s}_{bcng}(q+k'')I^{s}_{m'hed}(q)\right\}G_{gh}(k'-q).
\end{aligned}
\tag{2.60}
$$

The $\hat{\Lambda}$ is three-point vertex function defined by

$$
\Lambda_{abcdef}(q,q') = -T\sum_{p} G_{ab}(p+q)G_{cd}(p-q')G_{ef}(p). \tag{2.61}
$$

References

1. A.P. Mackenzie, Y. Maeno, Rev. Mod. Phys. **75**, 657 (2003)
2. Y. Maeno, S. Kittaka, T. Nomura, S. Yonezawa, K. Ishida, J. Phys. Soc. Jpn. **81** (2012)
3. M. Sigrist, Prog. Theor. Phys. Suppl. **160**, 1 (2005)
4. K. Ishida, H. Mukuda, Y. Kitaoka, K. Asayama, Z.Q. Mao, Y. Mori, Y. Maeno, Nature **396**, 658 (1998)
5. T. Nomura, K. Yamada, J. Phys. Soc. Jpn. **69**, 3678 (2000); T. Nomura, K. Yamada, J. Phys. Soc. Jpn. **71**, 1993 (2002)
6. Q.H. Wang, C. Platt, Y. Yang, C. Honerkamp, F.C. Zhang, W. Hanke, T.M. Rice, R. Thomale, Europhys. Lett. **104**, 17013 (2013)
7. S. Raghu, A. Kapitulnik, S.A. Kivelson, Phys. Rev. Lett. **105**, (2010)
8. T. Scaffidi, J.C. Romers, S.H. Simon, Phys. Rev. B **89** (2014)
9. I.A. Firmo, S. Lederer, C. Lupien, A.P. Mackenzie, J.C. Davis, S.A. Kivelson, Phys. Rev. B **88** (2013)
10. T. Takimoto, Phys. Rev. B **62**, R14641(R) (2000)
11. R. Tazai, Y. Yamakawa, T. Tsuchiisu, H. Kontani, Phys. Rev. B **94** (2016)
12. N.E. Bickers, S.R. White, Phys. Rev. B **43**, 8044 (1991)
13. M. Tsuchiizu, Y. Ohno, S. Onari, H. Kontani, Phys. Rev. Lett. **111** (2013)
14. M. Tsuchiizu, Y. Yamakawa, S. Onari, Y. Ohno, H. Kontani, Phys. Rev. B **91** (2015)
15. S. Onari, H. Kontani, Phys. Rev. Lett. **109** (2012)
16. S. Onari, Y. Yamakawa, H. Kontani, Phys. Rev. Lett. **112** (2014)

17. Y. Yamakawa, S. Onari, H. Kontani, Phys. Rev. X **6** (2016)
18. W. Metzner, M. Salmhofer, C. Honerkamp, V. Meden, K. Schönhammer, Rev. Mod. Phys. **84**, 299 (2012)
19. C.J. Halboth, W. Metzner, Phys. Rev. Lett. **85**, 5162 (2000)
20. C. Honerkamp, M. Salmhofer, Phys. Rev. Lett. **87**, 187004 (2001); C. Honerkamp, M. Salmhofer, Phys. Rev. B **64**, 184516 (2001)
21. A.A. Abrikosov, L.P. Gorkov, I.E. Dzyaloshinskii, *Methods of Quantum Field Theory in Statistical Physics* (Dover, New-York, 1975)
22. M. Braden, Y. Sidis, P. Bourges, P. Pfeuty, J. Kulda, Z. Mao, Y. Maeno, Phys. Rev. B **66** (2002)
23. J. Reiss, D. Rohe, W. Metzner, Phys. Rev. B **75** (2007)
24. J. Wang, A. Eberlein, W. Metzner, Phys. Rev. B **89**, 121116(R) (2014)

Chapter 3
Cooperation of el-ph and AFM Fluctuations

3.1 Background

3.1.1 el-ph Versus AFM Fluctuation

Recently, high T_c s-wave phase was observed in heavily electron-doped FeSe [1, 2] and A_3C_{60} (A =K, Rb, Cs) [3]. The schematic picture of the typical phase diagram is written in Fig. 3.1. Instead of the previously established common sense, s-wave SC phase appears even near the AFM phase. In principle, el-ph interaction should be important in such metals as discussed in some previous researches in Refs. [4–10]. However, the microscopic origin of the s-wave SC is still unsolved problem [11–13] (Figs. 3.1 and 3.2).

Common property of these s-wave SC states near AFM phase is orbital degrees of freedom. Therefore, it is naturally expected that s-wave SC can be induced by multi-orbital nature of the system as well as el-ph interaction. Then, we have to solve the following primitive question:

- Why s-wave Cooper pairs survive against strong AFM fluctuations?
- How the multi-orbital nature affects the s-wave SC state?

In the present study, we analyze multi-orbital Hubbard-Holstein (HH) models by considering the vertex correction and el-ph interaction as well as AFM fluctuations. As a result, we discover that s-wave SC phase is stabilized due to the cooperation of el-ph interaction and AFM fluctuation. This counter-intuitive mechanism is originating from the vertex correction, called χ-VC and U-VC. Also, we reveal that the "multi-orbital screening effect" is significant for explaining the s_{++}-wave SC phase.

R. Tazai, *Theory of Multipole Fluctuation Mediated Superconductivity and Multipole Phase*, Springer Theses,
https://doi.org/10.1007/978-981-16-1026-4_3

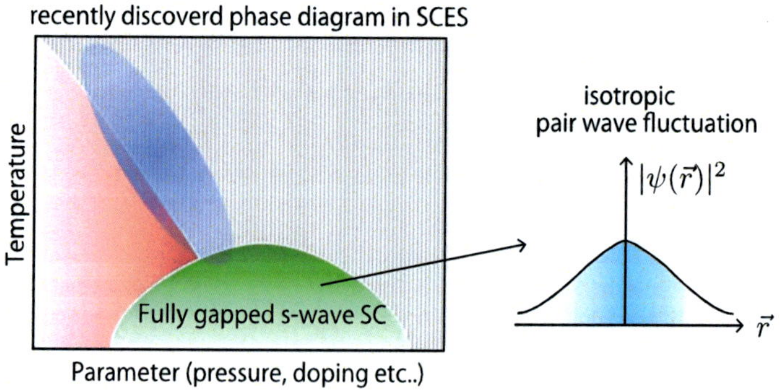

Fig. 3.1 Typical phase diagram recently discovered in multi-orbital system

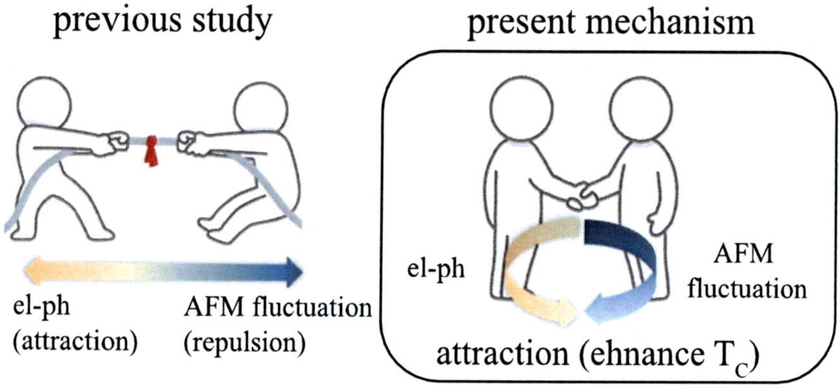

Fig. 3.2 (left) Competition between AFM fluctuation and el-ph interaction in the previous study. (right) The present proposed s-wave SC mechanism due to the cooperation among them

3.2 Formalism

3.2.1 Hubbard-Holstein Model (B_{1g} Phonon)

Kinetic Term

Here, we start from two-orbital Hubbard-Holstein model on 2D square lattice

$$\hat{H} = \hat{H}_{el} + \hat{H}_{\rm ph},$$
$$\hat{H}_{el} = \sum_{k\sigma lm} \xi_{k}^{lm} d_{kl\sigma}^{\dagger} d_{km\sigma} + \frac{1}{4}\sum_{i}\sum_{ll'mm'}\sum_{\sigma\sigma'\rho\rho'} U_{ll'mm'}^{\sigma\sigma'\rho\rho'} d_{il\sigma}^{\dagger} d_{il'\sigma'} d_{im\rho} d_{im'\rho'}^{\dagger}, \quad (3.1)$$

where $\hat{H}_{el}$ is composed of the kinetic and on-site Coulomb interaction term, i denotes index for lattice site, $d_{kl\sigma}^{\dagger}$ ($d_{kl\sigma}$) is the creation (annihilation) operator for d_{xz} or d_{yz}-electrons with wave number $\boldsymbol{k}$, orbital l, spin σ. $l, m = 1$ (2) corresponds to orbital index for $d_{xz}(d_{yz})$. We comment that our present results originate from multi-orbital nature of the electron system. Therefore, the result will change when we use one-orbital model as shown in Appendix 2. $\xi_{\boldsymbol{k}}^{lm}$ is defined as

$$\begin{aligned}\xi_{\boldsymbol{k}}^{11} &= -2t\cos k_x - 2t^{''}\cos k_y,\\ \xi_{\boldsymbol{k}}^{22} &= -2t\cos k_y - 2t^{''}\cos k_x,\\ \xi_{\boldsymbol{k}}^{12} = \xi_{\boldsymbol{k}}^{21} &= -4t^{'}\sin k_x \sin k_y.\end{aligned} \tag{3.2}$$

After that, the hopping parameters are fixed at $(t, t^{'}, t^{''}) = (1, 0.1, 0.1)$. Energy unit is $t = 1$, and the filling of d-electron takes $n_d = 2.30$. Obtained Fermi surfaces, so-called Fermi surface α and β, are given in Fig. 3.6a. θ is angle parameter for $\boldsymbol{k}$ on each Fermi surface. Multi-orbital on-site Coulomb interaction $\hat{U}$ includes intra-orbital U, inter-orbital U', Hund's coupling J, and pair hopping J' [15]. Spin-dependent bare four-point vertex $\hat{U}^{\sigma\sigma'\rho\rho'}$ is uniquely decomposed into spin and charge channels as

$$\hat{U}^{\sigma\sigma'\rho\rho'} = \frac{1}{2}\hat{U}^s \boldsymbol{\sigma}_{\sigma\sigma'} \cdot \boldsymbol{\sigma}_{\rho'\rho} + \frac{1}{2}\hat{U}^c \delta_{\sigma,\sigma'}\delta_{\rho',\rho}. \tag{3.3}$$

$\hat{U}^{s(c)}$ denotes spin (charge) channel of four-point vertex function. The detailed explanation for the Coulomb interaction was explained in the previous chapter.

Electron-Phonon (el-ph) Interaction

In the present study, we consider the effective el-ph interaction in B_{1g} symmetry, which is given by

$$\hat{H}_{\text{ph}} = \omega_D \sum_i b_i^\dagger b_i + \eta \sum_i (b_i^\dagger + b_i)(\hat{n}_i^{xz} - \hat{n}_i^{yz}), \tag{3.4}$$

where $\hat{n}_i^l$ is quantum operator for electron numbers with orbital l. Phonon creation (annihilation) operator is written by $b_i^\dagger$ (b_i). Coefficient η denotes coupling constant between d-electrons and B_{1g}-phonon. ω_D is Debye frequency of phonon.

- We analyze $d_x z$, $d_y z$ Hubbard-Holstein model.
- We consider electron-phonon interaction due to the B_{1g} phonon.

3.2.2 *Phonon-Mediated Electric Interaction*

Based on the present Hubbard-Holstein Hamiltonian, retarded interaction via el-ph coupling is given by

$$V = -g(\omega_j)\sum_i (\hat{n}_i^{xz} - \hat{n}_i^{yz})(\hat{n}_i^{xz} - \hat{n}_i^{yz}) \quad \left(g(\omega_j) = g\frac{\omega_D^2}{\omega_D^2 + \omega_j^2} \right), \tag{3.5}$$

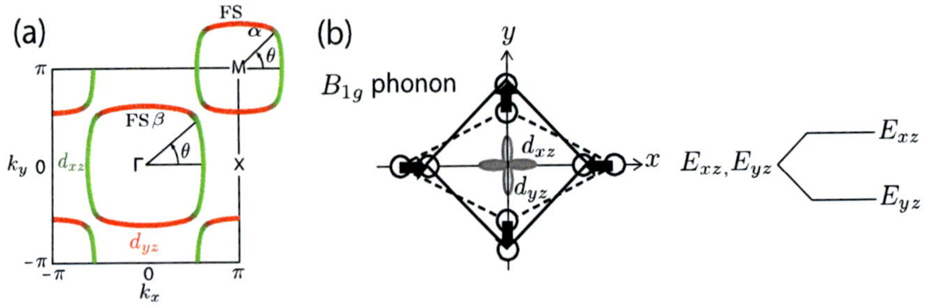

Fig. 3.3 **a** Obtained Fermi surfaces for $n_d = 2.3$. It is quite similar to that in Chap. 2. **b** The B_{1g}-type electron-phonon coupling caused by in-plane distortion. Reprinted with permission from Ref. [14]. Copyright ©2017 by the Physical Society of Japan

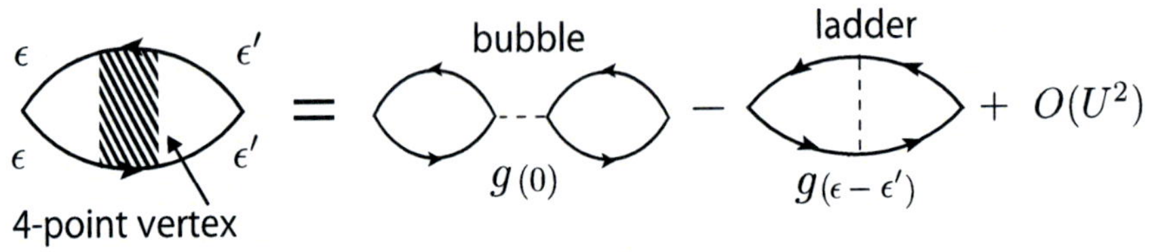

Fig. 3.4 The diagrammatic expression for the first-order correction included in p-h fluctuations

where $g \equiv \frac{2\eta^2}{\omega_D}$ (> 0) and $\omega_j = 2j\pi T$ is the Boson Matsubara frequency. We show schematic expression of possible B_{1g} phonon in Fig.3.3(b). Note that $n_{xz} \neq n_{yz}$ holds if the B_{1g}-orbital order occur.

Next, we derive the matrix elements of four-point vertex due to the B_{1g} phonon included in the p-h excitation. Figure 3.4 shows the first-order correction for p-h excitation process with zero exciting energy $\omega = 0$. In Fig.3.5, diagrammatic definition of the phonon-mediated interaction $\hat{C}_g^{\sigma\sigma\rho\rho}$ is shown. If we consider the condition $\omega_D \gg W_{\rm band}$, we obtain

$$C^s_{g,ll'mm'} = \begin{cases} g & (l = l' = m = m') \\ -g & (l = m \neq l' = m') \\ 0 & (\text{otherwise}), \end{cases} \tag{3.6}$$

$$C^c_{g,llmm'} = \begin{cases} +g & (l = l' = m = m') \\ -2g & (l = l' \neq m = m') \\ +g & (l = m \neq l' = m') \\ 0 & (\text{otherwise}), \end{cases} \tag{3.7}$$

where $\hat{C}_g^{c(s)} \equiv \hat{C}_g^{\uparrow\uparrow\uparrow\uparrow} + (-)\hat{C}_g^{\uparrow\uparrow\downarrow\downarrow}$. This spin dependence originates from the Pauli exclusion principle and Eq. (3.5). Therefore, $g(\omega_j)$ is replaced with the constant value g. In this case, both bubble and ladder diagrams contribute to p-h fluctuations.

On the other hand, if we consider the opposite situation $\omega_D \ll W_{\rm band}$, we obtain

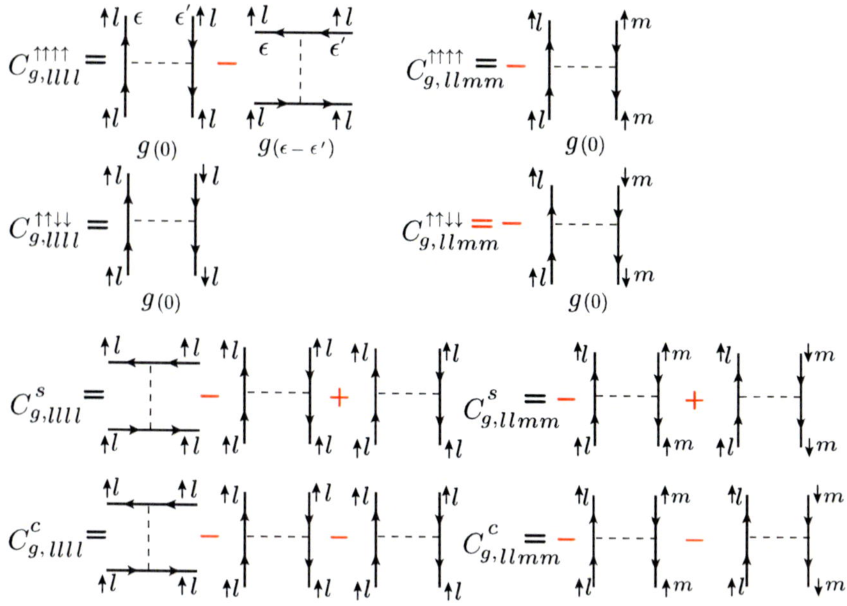

Fig. 3.5 The diagrammatic expression for the bare four-point vertex function due to the B_{1g}-type el-ph coupling. Here, we put $l \neq m$

$$C^s_{g,ll'mm'} = 0 \qquad \text{(all element)}, \tag{3.8}$$

$$C^c_{g,ll'mm'} = \begin{cases} +2g & (l = l' = m = m') \\ -2g & (l = l' \neq m = m') \\ 0 & \text{(otherwise)}. \end{cases} \tag{3.9}$$

Here, the ladder-type contribution is neglected based on the following reason. In the case of $\omega_D \ll W_{\text{band}}$, p-h excitation due to ladder term is

$$\chi^0(q) - T \sum_k G_{lm}(k+q) G_{m'l'}(k) \theta(\omega_D - |\epsilon|). \tag{3.10}$$

This contribution must be smaller than that without energy cutoff reduced by $\theta(\omega_D - |\epsilon|)$. Therefore, phonon-mediated four-point vertex under the condition of $\omega_D \ll W_{\text{band}}$ is rewritten by

$$C^c_{g,ll'mm'} = -2g(2\delta_{l,m} - 1)\delta_{l,l'}\delta_{m,m'} \tag{3.11}$$

Thus, four-point vertex $C^{s(c)}$ for spin (charge) channel due to the el-ph end Coulomb interaction is given by

$$\hat{C}^s \equiv \hat{U}^s \qquad \hat{C}^c \equiv \hat{U}^c - \hat{C}_g. \tag{3.12}$$

As a result, spin-channel four-point vertex is independent of the g for $\omega_D \ll W_{\rm band}$.

- We consider electric interaction induced by B_{1g} phonon.
- In the case of $\omega_D \ll W_{\rm band}$, phonon-mediated interaction works as charge-channel interaction.

3.2.3 Susceptibility with Self-Consistent Vertex Correction

Here, we introduce spin and charge susceptibilities considering χ-VC by performing self-consistent VC (SC-VC) method. Within SC-VC scheme, spin (charge) susceptibilities are given by

$$\hat{\chi}^{s(c)}(q) = \hat{\Phi}^{s(c)}(q)(\hat{1} - \hat{C}^s(c)\hat{\Phi}^{s(c)}(q))^{-1}, \tag{3.13}$$
$$\hat{\Phi}^{s(c)}(q) \equiv \hat{\chi}^0(q) + \hat{X}^{s(c)}(q)$$

The irreducible susceptibility $\hat{\chi}^0$ is calculated by

$$\chi^0_{ll'mm'}(q) = -T\sum_k G_{lm}(k+q)G_{m'l'}(k), \tag{3.14}$$

where Green function $G_{lm}(k)$ is in the orbital basis without self-energy effects Σ. $\hat{X}^{s(c)}(q)$ is vertex correction due to the χ-VC induced by the AL process introduced in Eq. (2.51) in the previous section. In the present SC-VC study, we obtain $\hat{X}^c(q)$ self-consistently, while $\hat{X}^s(q)$ is neglected since it is negligibly small as verified in Chap. 2. Here, spin (charge) Stoner factor $\alpha_{S(C)}$ is defined as the largest eigenvalue obtained from the following equation:

$$\left\{\hat{C}^{s(c)}\hat{\Phi}^{s(c)}(q)\right\}\mathbf{v} = \lambda\mathbf{v}. \tag{3.15}$$

3.3 Results

Hereafter, we show the numerical results of the present SC-VC study. The parameters are fixed at $J/U = 0.08$, $T = 5\times10^{-2}$, $(U, g) = (2.1, 0.15)$, and $(\alpha_S, \alpha_C) = (0.92, 0.93)$. In addition, 32×32 $\boldsymbol{k}$-meshes and 256 Matsubara frequencies are adopted.

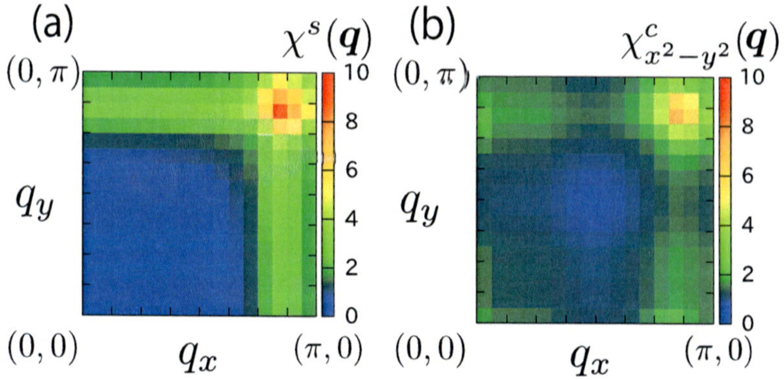

Fig. 3.6 **a** Total spin susceptibility $\chi^s(\boldsymbol{q})$. **b** Orbital susceptibility $\chi^c_{x^2-y^2}(\boldsymbol{q})$ belonging to B_{1g}-symmetry at $(\alpha_S, \alpha_C) = (0.92, 0.93)$ and $(U, g) = (2.1, 0.15)$. Reprinted with permission from Ref. [14]. Copyright ©2017 by the Physical Society of Japan

3.3.1 B_{1g}-Orbital Fluctuations

Figure 3.6a, b shows obtained total spin susceptibility χ^s and B_{1g} orbital susceptibility $\chi^c_{x^2-y^2}$ calculated by SC-VC method. The definitions of them are given by

$$\chi^s(\boldsymbol{q}) = \sum_{lm} \chi^s_{llmm}(\boldsymbol{q})$$
$$\chi^c_{x^2-y^2}(\boldsymbol{q}) = \sum_{lm} (-1)^{l+m} \chi^c_{llmm}(\boldsymbol{q}). \tag{3.16}$$

Note that the orbital-ordered phase with order parameter $\hat{n}_{xz} - \hat{n}_{yz}$ will appear if $\chi^c_{x^2-y^2}$ diverges. As a result, we find that B_{1g} orbital susceptibility has large value almost equivalent to spin susceptibility. Then, $\chi^s \approx \chi^c$ is realized even in the presence of quite small $g \approx 0.15$ ($\lesssim 0.1U$). This result is understood by following approximate equation:

$$\chi^c_{x^2-y^2}(\boldsymbol{Q}) \sim \Phi^c(\boldsymbol{Q})[1 - (2U' - U + 4g)\Phi^c(\boldsymbol{Q})]^{-1}, \qquad \left(\alpha_C \lesssim 1\right) \tag{3.17}$$

where $\Phi^c(\boldsymbol{Q})$ corresponds to irreducible susceptibility for intra-orbital channel. In the presence of U', charge-channel Stoner factor is written by

$$\alpha_C = (2U' - U + 4g)\Phi^c(\boldsymbol{Q}). \tag{3.18}$$

Thus, α_C reaches unity in the presence of small B_{1g}-phonon. α_C is further enhanced by the AL-type χ-VC: $\Phi^c(\boldsymbol{Q}) \gg \chi^0(\boldsymbol{Q})$. In this case, strong orbital fluctuations appear due to the cooperation of B_{1g}-phonon and χ-VC. In contrast, we need large $g \ll U$ to obtain χ^c without χ-VC. We comment that charge-channel χ-VC at $q = 0$ is written as

$$\Phi^c(q=0) \propto T \sum_p \{3\chi^s(p)^2 + \chi^c(p)^2\}, \tag{3.19}$$

where the effect of g in $\chi^c(p)^2$ is neglected since. The contribution from $\chi^s(q)^2$ overwhelms that from $\chi^c(q)^2$. Therefore, we safely put $g = 0$ in χ-VC.

- Strong orbital fluctuations χ^c are induced by the cooperation of small B_{1g} phonon $g \lesssim U/10$ and spin susceptibility due to the significant roles of χ-VC.

3.3.2 Superconductivity

SC Pairing Interaction

Now, we analyze SC state beyond Migdal-Eliashberg (ME) scheme. The spin singlet SC pairing interaction is given by

$$\hat{V}(k,k') = \frac{3}{2}\hat{V}^s(k,k') - \frac{1}{2}\hat{V}^c(k,k') - \hat{C}^s. \tag{3.20}$$

The pairing interaction includes three-point vertex correction $\hat{\Lambda}$ due to the χ-VC,

$$\hat{V}^{s(c)}(k,k') = \hat{\Lambda}^{s(c)}(k,k')\left\{\hat{C}^{s(c)}\hat{\chi}^{s(c)}(k-k')\hat{C}^{s(c)} + \hat{C}^{s(c)}\right\}\hat{\Lambda}^{s(c)*}(-k,-k'), \tag{3.21}$$

where $\Lambda^{s(c)*}_{ll'mm'} \equiv \Lambda^{s(c)}_{m'ml'l}$. $\hat{\Lambda}^{s(c)}$ is the AL-type vertex function, which we call U-VC. We set $g = 0$ in the U-VC since the effect of g is negligibly small in the same way to χ-VC. In Fig. 3.7a, b, charge- and spin-channel U-VC are plotted at $\epsilon_n = \epsilon_{n'} = \pi T$. We obtain large enhancement factor for the charge-channel $|\Lambda^c|^2 \gg 1$, which is understood by AL-type U-VC given by

$$\Lambda^{\mathrm{AL},c}(q) \propto \sum_p \chi^s(p)\chi^s(p+q). \tag{3.22}$$

In contrast, $|\Lambda^s|^2 \ll 1$ is obtained for the spin channel. α_S-dependent of the enhancement factor is plotted in Fig. 3.8(a). The diagrammatic expression of the AL-term is shown in Fig. 3.8(b). As a result, we obtain large enhancement factor for the charge-channel attractive interaction by considering the U-VC (Fig. 3.8).

SC Phase Diagram

Now, we solve the gap equation given by

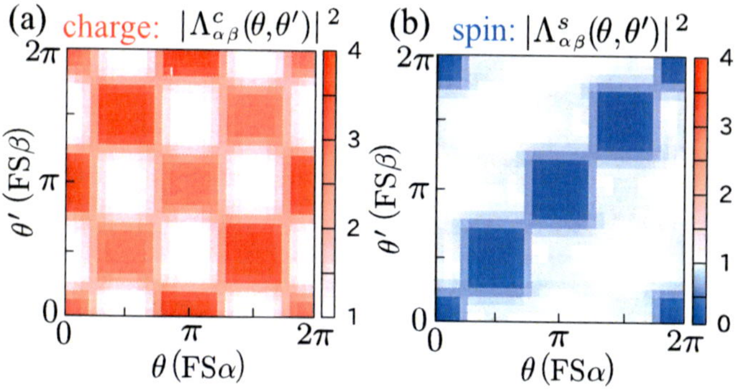

Fig. 3.7 **a, b** U-VC for charge channel and spin channel. θ (θ') is angle of the Fermi point on the Fermi surface. Reprinted with permission from Ref. [14]. Copyright ©2017 by the Physical Society of Japan

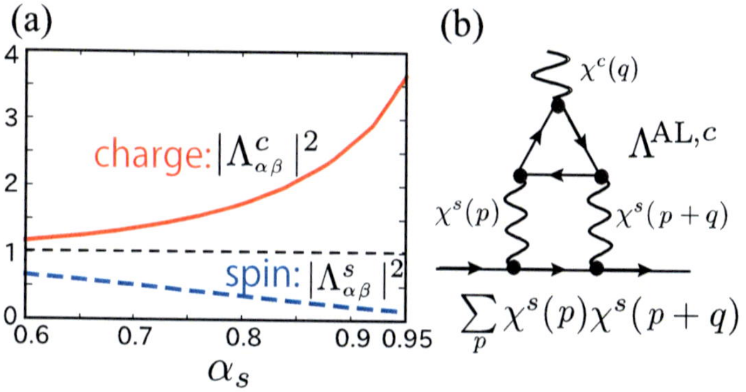

Fig. 3.8 **a** Obtained U-VC as a function of α_S at $\theta = \theta' = 0$. **b** Feynman diagram of AL-term for charge channel. Reprinted with permission from Ref. [14]. Copyright ©2017 by the Physical Society of Japan

$$\lambda\Delta^a(\theta,\epsilon_n) = -\frac{\pi T}{(2\pi)^2}\sum_{a'\epsilon_m}\int_0^{2\pi}\frac{d\theta'}{v_{a'\theta'}}\left|\frac{\partial \boldsymbol{k}_{a'\theta'}}{\partial\theta'}\right|\frac{\Delta^{a'}(\theta',\epsilon_m)}{|\epsilon_m|}V^{aa'}(\theta\epsilon_n,\theta'\epsilon_m), \quad (3.23)$$

where $\Delta^a(\theta,\epsilon_n)$ is spin singlet gap function. Index a, a' denote band index and λ is eigenvalue of the gap equation. $v_{a\theta}$ is the Fermi velocity. $V^{aa'}$ is SC interaction in the band basis, calculated by unitary transformation of $\hat{V}^{s(c)}(k,k')$.

In Fig. 3.9a, we show SC phase diagram from the linearized gap equation. We obtain three types of s-wave states as given in Fig. 3.9b–d As a result, fully gapped s_{++} state without any sign reversal is realized for a wide region around $\alpha_C \sim \alpha_S \geq 0.8$, which comes from enlarged attraction due to $|\Lambda^c|^2$. In contrast, as we show in Fig. 3.9e, the full gap s_{++} state disappears in the absence of U-VC. Then, we obtain

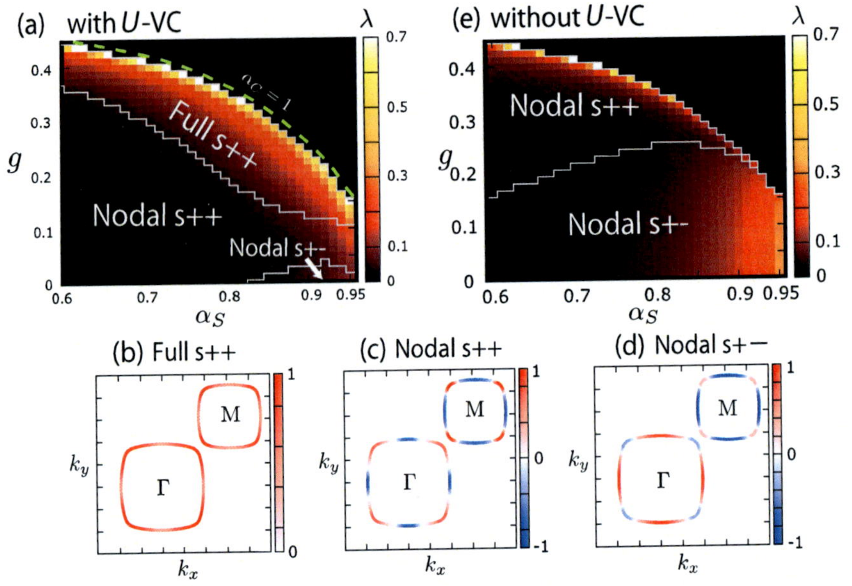

Fig. 3.9 **a** SC phase diagram calculated by the gap equation with U-VC. Gap structures for **b** s_{++}, **c** nodal s_{++}, and **d** nodal s_{+-} states. **e** Phase diagram in the absence of U-VC. Reprinted with permission from Ref. [14]. Copyright ©2017 by the Physical Society of Japan

$$\Delta = s_{++}\text{-wave SC} \quad (\text{large } \alpha_S),$$
$$\Delta = \text{nodal}s\text{-wave SC} \quad (\text{small } \alpha_S). \tag{3.24}$$

Thus, significant roles of U-VC for s-wave pairing mechanism is clearly confirmed. This fact leads to the violation of Migdal-Eliashberg (ME) theorem due to the strong $\boldsymbol{q}$-dependence of SC pairing interaction As a result, s_{++} wave state is stabilized by large attraction due to the charge-channel AL-type U-VC:

$$V^c \simeq \frac{1}{2}\left\{U - 4g + (2U' - U + 4g)^2 \chi^c_{x^2-y^2}(\boldsymbol{Q})\right\} |\Lambda^c|^2. \tag{3.25}$$

Multi-orbital Screening Effect

Here, we show the important roles of "multi-orbital screening effect" in the multi-orbital systems, by which the effect of on-site Coulomb repulsion for s-wave SC is reduced. In Fig. 3.10a, b, depairing processes for intra-orbital Cooper pairs are plotted in the case of one- and two-orbital systems, respectively. In one-orbital system, the energy loss due to the intra-orbital Copper pairs is given by U, while it is reduced to $\sim (U - U')$ in multi-orbital system. Then, we reveal that

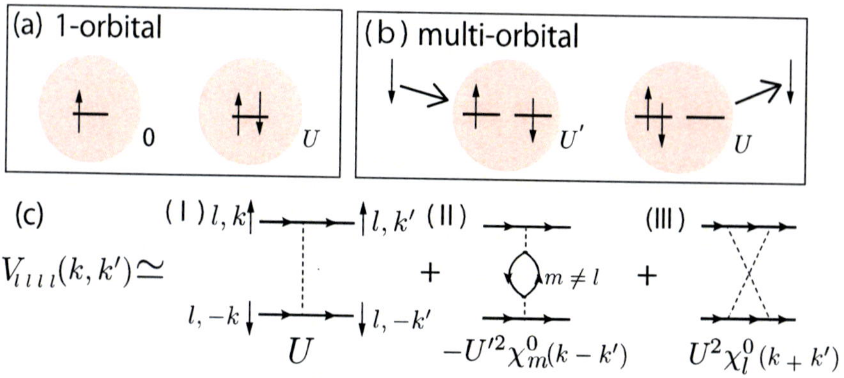

Fig. 3.10 Depairing processes of intra-orbital Cooper pairs in **a** one-orbital and **b** two-orbital systems. In multi-orbital system, energy loss is reduced by "multi-orbital screening effect". **c** Pairing interaction up to the second order of U. The process (II) appears only in the multi-orbital systems. Reprinted with permission from Ref. [14]. Copyright ©2017 by the Physical Society of Japan

$$\begin{aligned} &\text{energy loss} = U \quad \text{(one-orbital)},\\ &\text{energy loss} = U - U' \quad \text{(multi-orbital)}. \end{aligned} \tag{3.26}$$

This reduction is significant only in multi-orbital system as shown in Fig. 3.10b. In particular, the most significant contribution comes from second order of U in the SC pairing interaction, which is given by

$$V_{llll}^{\uparrow\downarrow\uparrow\downarrow}(k,k') \propto \sum_m U - U'^2\chi^0_{mmmm}(k-k') + U^2\chi^0_{llll}(k+k') \quad (l \neq m). \tag{3.27}$$

The diagrammatic expression is shown by (II) in Fig. 3.10c. Here, we assume $G_{lm} = G_l\delta_{l,m}$ for simplicity. Therefore, screening effect becomes prominent when $U'\chi^0_m \sim O(1)$ for $l \neq m$. In conclusion, the obtained results are summarized as follows:

- s_{++}-wave SC phase appears near the AFM-QCP due to U-VC and χ-VC.
- Origin of s_{++}-wave state is cooperation between B_{1g} phonon and AFN fluctuation.
- Multi-orbital screening effects play important roles for s-wave Cooper pairs.

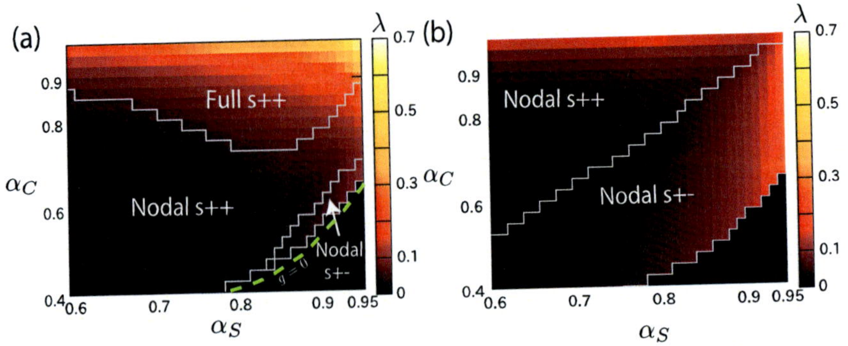

Fig. 3.11 **a** SC phase diagram by considering the momentum dependence of U-VC. **b** Phase diagram based on the local approximation for U-VC. Reprinted with permission from Ref. [14]. Copyright ©2017 by the Physical Society of Japan

3.3.3 *Momentum Dependence of U-VC*

Here, we explain the importance of the momentum dependence of U-VC for s_{++}-wave SC state. In Fig. 3.11a, we show the SC phase diagram in α_S α_C space, in which momentum dependence of U-VC is considered. On the other hand, Fig. 3.11b shows the SC phase diagram based on local approximation, in which momentum dependence of U-VC is neglected. Based on the local approximation, U-VC is given by

$$\hat{\Lambda}^{s(c)}_{\mathrm{loc}}(\epsilon_n, \epsilon_{n'}) = \langle \hat{\Lambda}^{s(c)}(k, k') \rangle_{\boldsymbol{k},\boldsymbol{k}'}. \tag{3.28}$$

We find that s_{++}-wave phase disappears in this case. In particular, obtained phase diagram is quite similar to that when U-VC is neglected as shown in Fig. 3.9e. Therefore, we conclude that the momentum dependence of U-VC is significant in terms of s_{++}-wave SC paring mechanism.

- Momentum dependence of U-VC is important for s_{++}-wave SC phase.

3.3.4 *Filling Dependence*

Here, we discuss filling dependence of the SC phase diagram. Figure 3.12 shows obtained SC phase diagram as a function of chemical potential μ and el-ph interaction g. Note that $n_d = 2.3$ in the previous sections corresponds to $\mu = 0.5$. Figure 3.12a shows the phase diagram by considering the U-VC. Various SC states including

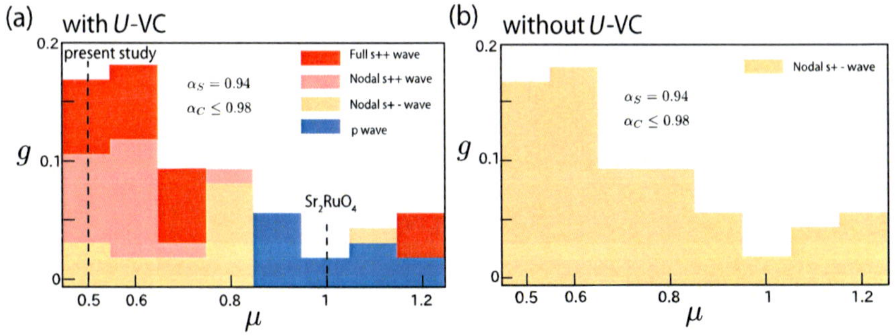

Fig. 3.12 SC phase diagram by **a** considering or **b** neglecting the U-VC. U is set at $\alpha_S = 0.94$. The white-colored region is $\alpha_C > 0.98$. Reprinted with permission from Ref. [14]. Copyright ©2017 by the Physical Society of Japan

the s_{++}-wave phase appears for a wide range of filling parameter. It is noteworthy that TSC phase emerges around $\mu \simeq 1$, which is equivalent to the previous work by fRG+cRPA study (Chap. 2). On the other hand, s_{++}-wave SC phase disappears if U-VC is dropped as shown in Fig. 3.12b. Therefore, U-VC plays important roles for realizing the fully gaped s_{++} wave pairs for wide parameter region.

- U-VC brings rich variety of SC state, including TSC phase.
- s_{++}-wave SC state is strong since it appears in wide parameter region.

3.3.5 *Retardation and Impurity Effects*

Retardation Effects

Here, we consider retardation effect given by $\omega_D \ll T$, that is,

$$g(\omega_j) = g\delta_{j,0} \text{ (with retardation effects).} \tag{3.29}$$

The phase diagram considering the retardation effect is given in Fig. 3.13. It shows that s_{++}-wave SC region is drastically expanded compared with Fig. 3.11a. Therefore, the retardation effect is important to obtain s_{++}-wave SC state.

Impurity Effects

Next, we analyze the SC phase in the presence of dilute non-magnetic impurities based on T-matrix approximation. The linearized gap equation in the presence of impurity is given by

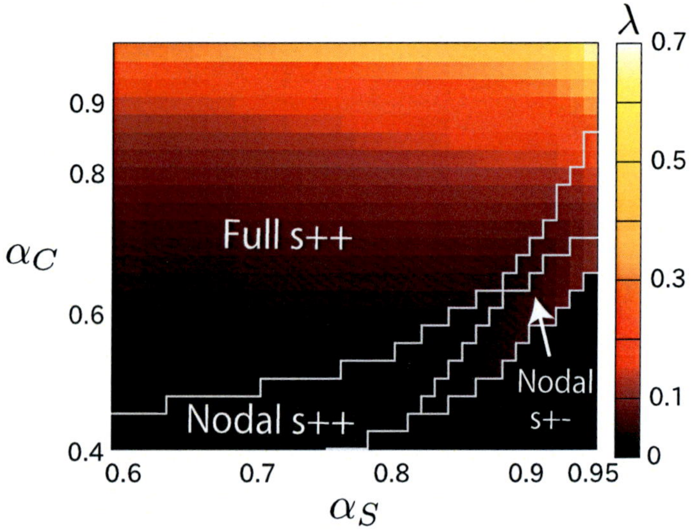

Fig. 3.13 SC phase diagram by considering the strong retardation effect. s_{++}-wave region is drastically expanded due to the retardation effect. Reprinted with permission from Ref. [14]. Copyright ©2017 by the Physical Society of Japan

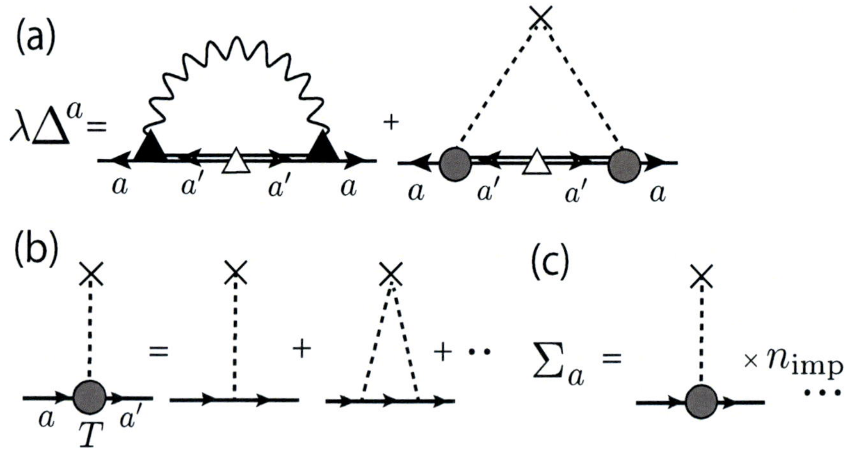

Fig. 3.14 **a** Linearized gap equation with impurities. **b** Diagrammatic expression of T-matrix due to the single impurity. **c** Self-energy included by impurity

$$\lambda\Delta^a(\boldsymbol{k},\epsilon_n) = \frac{-T}{(2\pi)^2}\sum_{a'\epsilon_m\boldsymbol{k}'}|G_{a'}(\boldsymbol{k}',\epsilon_m)|^2\Delta^{a'}(\boldsymbol{k}',\epsilon_m)$$

$$\times\left[V^{aa'}(\boldsymbol{k},\epsilon_n,\boldsymbol{k}',\epsilon_m) - \frac{n_{\text{imp}}}{T}|T_{aa'}(\boldsymbol{k},\boldsymbol{k}',\epsilon_m)|^2\delta_{n,m}\right], \quad (3.30)$$

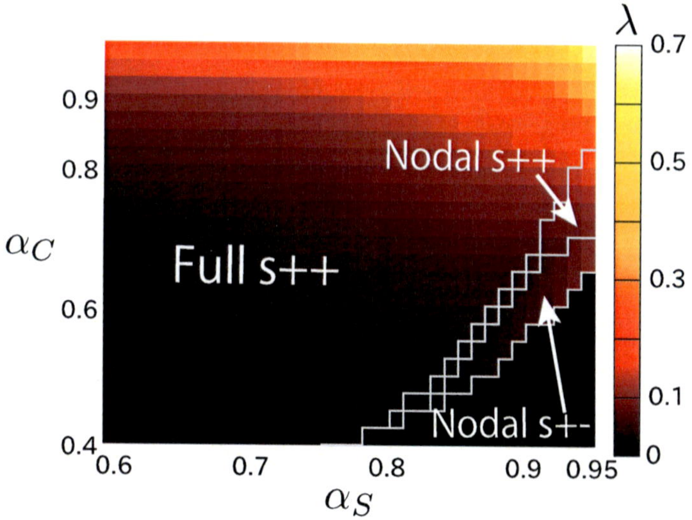

Fig. 3.15 SC phase diagram at $n_{\mathrm{imp}} = 0.1\%$ in $\alpha_S - \alpha_C$ space. s_{++} wave region is expanded by doping the impurities. Reprinted with permission from Ref. [14]. Copyright ©2017 by the Physical Society of Japan

which is diagrammatically shown in Fig. 3.14a. Here n_{imp} denotes concentration of the impurity and $T_{aa'}(\boldsymbol{k}, \boldsymbol{k}', \epsilon_m)$ is T-matrix due to the impurity as shown in Fig. 3.14b. Double line corresponds to the Green function with the impurity-induced self-energy, which is given by

$$\Sigma_a(k) = n_{\mathrm{imp}} T_{aa}(\boldsymbol{k}, \boldsymbol{k}, \epsilon_n). \tag{3.31}$$

The diagrammatic expression of the self-energy is shown in Fig. 3.14c.

Figure 3.15 shows the obtained SC phase diagram with impurity in the $\alpha_S - \alpha_C$ space. We find that the area of s_{++}-wave state is drastically expanded by the impurity effect even for $n_{\mathrm{imp}} = 0.1\%$.

- Retardation effect due to the B_{1g} phonon stabilizes s-wave SC state.
- Impurity effect enhances the region of s-wave SC phase even at $n_{\mathrm{imp}} = 0.1$.

3.4 Summary

In summary, we proposed the microscopic origin of the s_{++}-wave SC phase in multi-orbital systems. First, we demonstrated that orbital fluctuations drastically develop due to the cooperation of χ-VC and B_{1g} el-ph interaction. Then, orbital-fluctuation-mediated attractive force is enlarged by charge-channel U-VC, while repulsion due to the spin fluctuations is reduced by spin-channel U-VC. As a result, we obtain s_{++}-wave SC phase with large eigenvalue emerging near the AFM phase.

As the origin of the s_{++}-wave state, energy loss due to the on-site Coulomb repulsion is reduced by "multi-orbital screening effect". Moreover, the momentum dependence of U-VC is important to obtain s_{++}-wave SC phase. Note that charge-channel U-VC is also enlarged even in one-orbital system as shown in Appendix 2. Therefore, Kohn anomaly observed in high-T_c cuprate superconductor can be explained based on the present proposed theory.

The main findings of the present study are listed as follows:

- Based on the two-orbital Hubbard-Holstein model, we reveal that s_{++}-wave SC phase appears near AFM-QCP.
- Microscopic origin of s_{++}-wave state comes from cooperation between AFM fluctuations and el-ph interaction in B_{1g} symmetry due to the χ-VC and U-VC.
- U-VC brings rich variety of phase diagram including SSC and TSC phases.
- Retardation and impurity effects drastically stabilize s-wave SC phase.
- Energy loss due to the on-site Coulomb repulsion is suppressed by multi-orbital screening effect.
- Momentum dependence of U-VC is prominent for s-wave SC.

Appendix 2

One-orbital Electron Systems

Here, we analyze the one-orbital system to demonstrate that the χ-VC is important in one-orbital system as well as two-orbital system. Thus, cooperation between el-ph and AFM fluctuation can be important even in one-orbital electrons such as in cuprate superconductor. Here, we use the energy dispersion defined by

$$\xi_k = 2t(\cos k_x + \cos k_y) + 4t' \cos k_x \cos k_y + 2t''(\cos 2k_x + \cos 2k_y), \quad (3.32)$$

where $(t, t', t'') = (-1, 1/6, -1/5)$. The chemical potential is set at $\mu = -0.8$ in the following numerical study. Figure 3.16 shows band dispersion and Fermi surface.

First, we perform the RPA to obtain p-h susceptibility (Fig. 3.17). Obtained charge- and spin-channel susceptibility is plotted in Fig. 3.17a, b, respectively. Here we set $\alpha_S = 0.9$ and $g = 0$. Based on the RPA, spin-channel susceptibility is much larger than charge-channel one in the absence of g. Then, we consider el-ph interaction

Table 3.1 Elements of spin- and charge-ch four-point vertex function

	Bare four point	Bare four point	Single fluctuation
Spin channel	0	U	$U^2\chi^s_{RPA}$
Charge channel	g	$-U$	$U^2\chi^c_{RPA}$

in A_{1g}-symmetry. In this case, g is added to SC paring interaction in linearized gap equation. Here, we consider only single-fluctuation term in Fig. 2.13. In this case, the SC paring interaction is given by Table 3.1. Thus, the charge-channel susceptibility includes g, which is written by

$$\chi^c_{RPA} = \frac{\chi^0}{1 + (U - g)\chi^0}. \tag{3.33}$$

The charge-channel Stoner factor is fixed at $\alpha_C = 0.7$.

In addition, we consider AL-type U-VC for the SC paring interaction. In Fig. 3.18a, b, obtained s- and d-wave gap functions are plotted. s-wave gap is quite isotropic while d-wave one is anisotropic and belongs to $d_{x^2-y^2}$-symmetry. In Fig. 3.18c, effective SC paring interaction is plotted. d-wave SC state appears when the spin Stoner factor α_S is small while that for both s- and d-wave simultaneously increase as α_S becomes large. Thus, we conclude that attractive SC paring interaction for isotropic s-wave state is strongly enlarged by AL-VC even in one-orbital system as well as two-orbital one. Then, we obtain

$$V_d \gg V_s \text{ (small } a_S), \tag{3.34}$$

$$V_d \simeq V_s \text{ (near AFM-QCP)}. \tag{3.35}$$

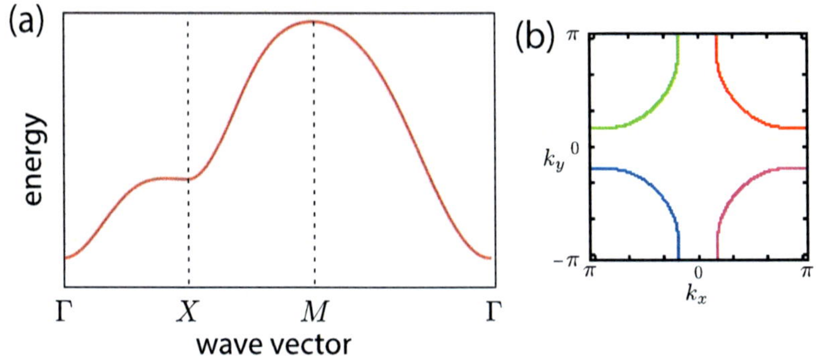

Fig. 3.16 **a** Obtained band dispersion and **b** Fermi surface based in the one-orbital tight-binding model

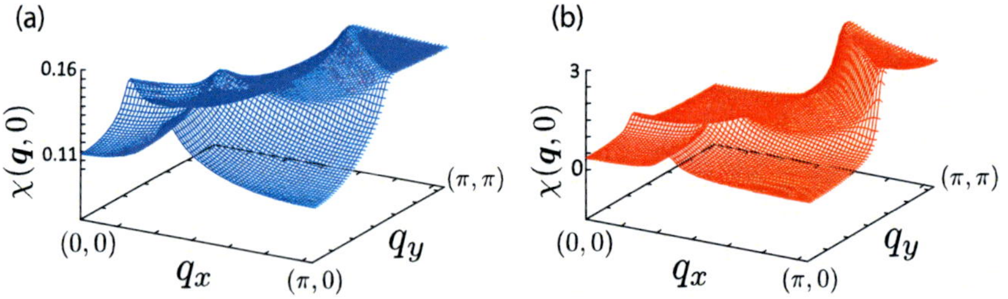

Fig. 3.17 Obtained susceptibility in **a** charge and **b** spin channel

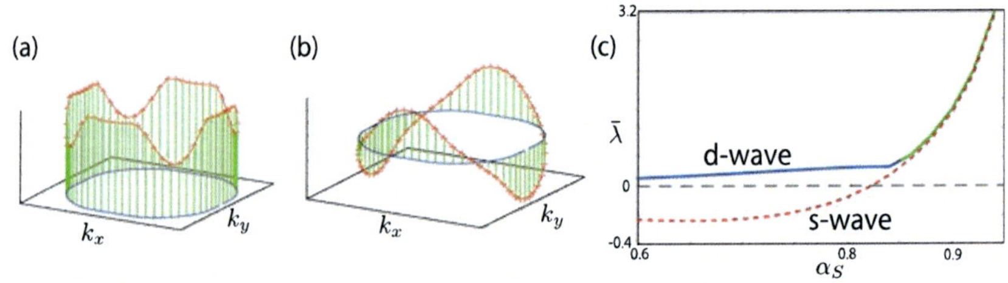

Fig. 3.18 Gap function on the Fermi surface for **a** s-wave and **b** $d_{x^2-y^2}$-wave states. **c** Effective SC paring interaction. Red dotted (blue solid) line corresponds to s- (d-) wave SC state

In Fig. 3.19a, b, we show obtained U-VC due to AL-term. Charge-channel U-VC is about ~ 3 while the maximum value of spin channel is only about $\sim$0.8. Therefore, AL-type U-VC strongly enhances the attractive paring interaction due to the charge fluctuations. On the other hand, only d-wave state is stabilized near AFM-QCP without U-VC as shown in Fig. 3.19c. In conclusion, we discover that U-VC enhances the attractive pairing interaction in one-orbital model. Thus,

$$V_d \gg V_s \ \text{(without } U\text{-VC)}, \tag{3.36}$$

$$V_d \ll V_s \ \text{(with } U\text{-VC)}. \tag{3.37}$$

Finally, we plot the effective interaction for isotropic gap function based on local approximation in Fig. 3.19d. The s-wave SC interaction is much larger than that in local approximation (about factor $\sim$3). This result is understood by the following relation:

$$U\text{-VC} \propto \Big(\sum_q \chi^s(q)\Big)^2 \propto \xi^0 \ \text{(local approx.)}, \tag{3.38}$$

$$U\text{-VC} \propto \sum_q (\chi^s(q))^2 \propto \xi^2 \ \text{(with } k\text{-dependence)}, \tag{3.39}$$

where ξ is correlation length. Therefore, it is verified that momentum dependence of U-VC is important in one-orbital system as well as two-orbital system.

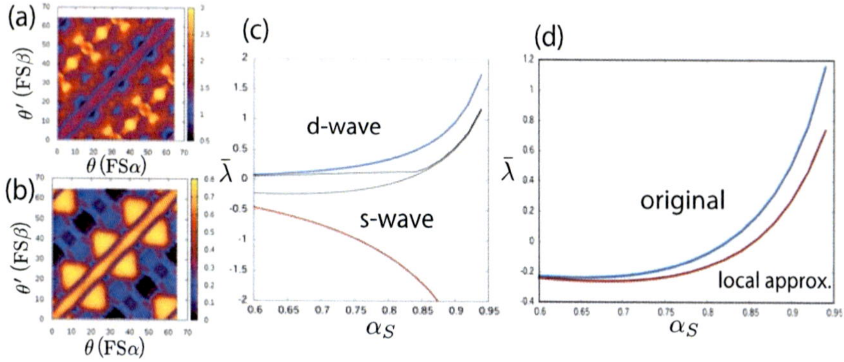

Fig. 3.19 **a** Charge-channel and **b** spin-channel enhancement factor given by $(U\text{-VC})^2$. **c** Effective paring interaction $\bar{\lambda}$ in the absence of U-VC. **d** $\bar{\lambda}$ based on the local approximation (red line) and original one (blue line)

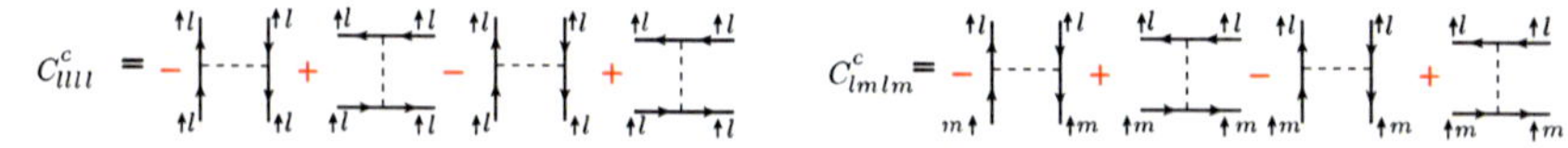

Coulomb	×	×	-U	0
el-ph (Pauli)	×	×	g(0)	0
el-ph	g(0)	-g(ε - ε′)	g(0)	0

Coulomb	-J	U′	-J	0
el-ph (Pauli)	0	g(ε - ε′)	0	0
el-ph	0	g(ε - ε′)	0	0

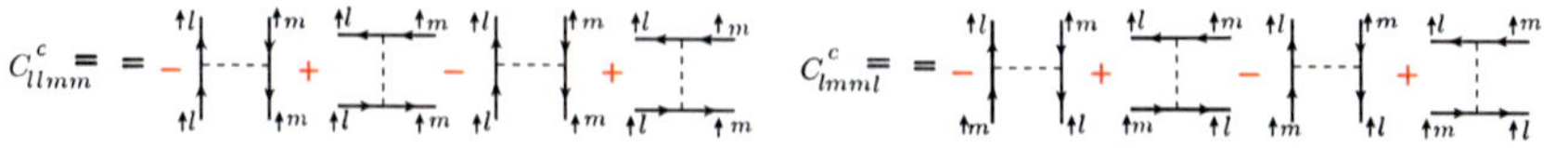

Coulomb	-U′	J	-U′	0
el-ph (Pauli)	-g(0)	0	-g(0)	0
el-ph	-g(0)	0	-g(0)	0

Coulomb	0	0	-J′	0
el-ph (Pauli)	0	0	0	0
el-ph	0	0	0	0

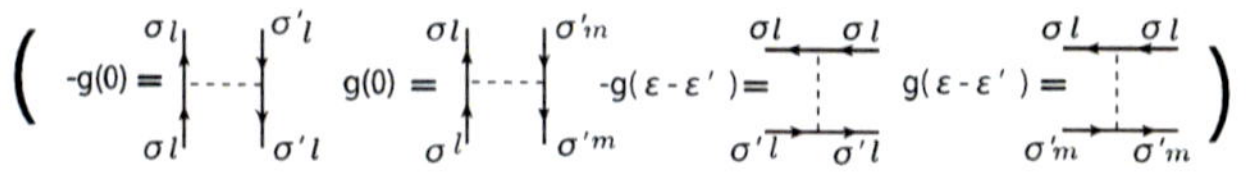

Fig. 3.20 Definition of the charge-channel bare four-point vertex due to B_{1g}-phonon in two-orbital (d_{xz}, d_{yz}) model

B_{1g} Phonon

Here, we explain detailed explanation of the B_{1g} phonon in two-orbital Hubbard-Holstein model. The charge- and spin-channel four-point vertex function is given in Figs. 3.20 and 3.21, respectively. Note that Pauli exclusion principle is considered for the on-site Coulomb interaction.

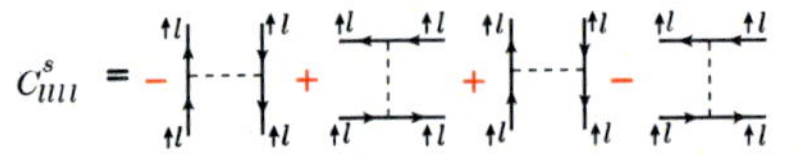

Coulomb	×	×	U	0
el-ph (Pauli)	×	×	-g(0)	0
el-ph	g(0)	-g(ε - ε')	-g(0)	0

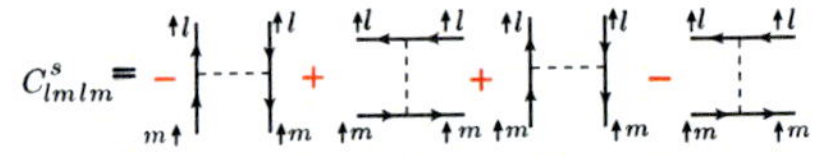

Coulomb	-J	U'	J	0
el-ph (Pauli)	0	g(ε - ε')	0	0
el-ph	0	g(ε - ε')	0	0

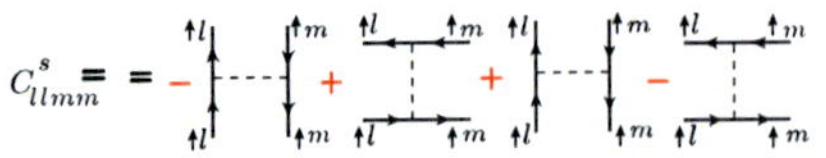

Coulomb	-U'	J	U'	0
el-ph (Pauli)	-g(0)	0	g(0)	0
el-ph	-g(0)	0	g(0)	0

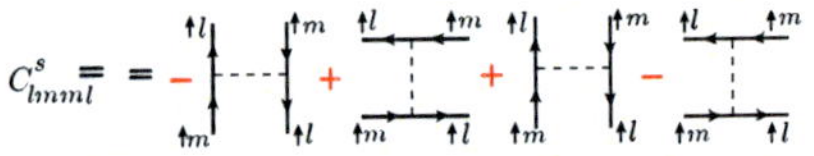

Coulomb	0	0	J'	0
el-ph (Pauli)	0	0	0	0
el-ph	0	0	0	0

Fig. 3.21 Definition of the spin-channel bare four-point vertex due to B_{1g}-phonon in two-orbital (d_{xz}, d_{yz}) model

References

1. C.H.P. Wen, H.C. Xu, C. Chen, Z.C. Huang, Y.J. Pu, Q. Song, B.P. Xie, M. Abdel-Hafiez, D.A. Chareev, A.N. Vasiliev, R. Peng, D.L. Feng, Nat. Commun. **7**, 10840 (2016)
2. Y.J. Yan, W.H. Zhang, M.Q. Ren, X. Liu, X.F. Lu, N.Z. Wang, X.H. Niu, Q. Fan, J. Miao, R. Tao, B.P. Xie, X.H. Chen, T. Zhang, D.L. Feng, Phys. Rev. B **94**, (2016)
3. Y. Takabayashi, L. Prassides, Philos. Trans. R. Soc. A **374**, 20150320 (2016)
4. S. Rebec, T. Jia, C. Zhang, M. Hashimoto, D. Lu, R. Moore, Z. Shen, arXiv:1606.09358
5. Y. Zhou, A.J. Millis, Phys. Rev. B **93** (2016)
6. S. Choi, W.-J. Jang, H.-J. Lee, J.M. Ok, H.W. Choi, A.T. Lee, A. Akbari, H. Suh, Y.K. Semertzidis, Y. Bang, J.S. Kim, J. Lee, arXiv:1608.00886
7. L. Rademaker, Y. Wang, T. Berlijn, S. Johnston, New J. Phys. **18** (2016)
8. M. Capone, M. Fabrizio, E. Tosatti, Phys. Rev. Lett. **86**, 5361 (2001)
9. Y. Nomural, S. Sakai, M. Capone, R. Arita, Sci. Adv. **1**, e1500568 (2015)
10. M. Kim, Y. Nomura, M. Ferrero, P. Seth, O. Parcollet, A. Georges, Phys. Rev. B **94**, (2016)
11. Z.B. Huang, W. Hanke, E. Arrigoni, D.J. Scalapino, Phys. Rev. B **68**, 220507(R) (2003)
12. O. Bodensiek, R. Zitko, M. Vojta, M. Jarrell, T. Pruschke, Phys. Rev. Lett. **110**, (2013)
13. K. Masuda, D. Yamamoto, Phys. Rev. B **91** (2015)
14. R. Tazai, Y. Yamakawa, M. Tsuchiizu, H. Kontani, J. Phys. Soc. Jpn. **86** (2017)
15. T. Takimoto, Phys. Rev. B **62**, R14641(R) (2000)

Part II
f-electron Systems

Chapter 4
Fully Gapped Superconductivity in $CeCu_2Si_2$

4.1 Background

In heavy fermion systems, exotic electronic phenomena are induced by mixing of strong Coulomb repulsion and spin-orbit coupling (SOI). Especially, higher rank multipole degrees of freedoms play significant roles due to the strong SOI in f-electrons. For instance, quadrupole (rank 2) and octupole (rank 3) phase transitions were reported in CeB_6 [8, 9]. Also, hexadecapole (rank 4) and dotriacontapole (rank 5) ordering were predicted in $PrRu_4P_{12}$ [10] and URu_2Si_2 [11–13].

In terms of superconductivity (SC), multipole degrees of freedoms cause exotic SC phase. For example, unconventional SC phase emerges next to the quadrupole phase in PrT_2Zn_{20} (T = Rh and Ir) [15] and PrT_2Al_{20} (T=V,Ti) [16]. Moreover, coexistence of SC phase and multipole phase was observed in URu_2Si_2. Recently, spin triplet SC phase is reported in UTe_2. These experiments indicate that various exotic SC phases originate from higher rank (≥ 2) multipole fluctuations.

In the present study, we focus on $CeCu_2Si_2$ well known as the first discovered heavy fermion superconductor [17–19]. SC transition occurs at $T_c \approx 0.6$K near the AFM quantum critical point at ambient pressure [20], while it goes up to 1.5K around $P_c \approx 4.5$GPa. Historically, it was believed as a typical nodal d-wave superconductor reflecting strong AFM fluctuations since nodal-like behavior was observed in thermodynamic quantity by NMR and specific heat measurements. The d-wave SC was explained by Migdal-Eliashberg approximation, in which nodal SC states are stabilized so as to avoid the energy loss due to the repulsion by AFM fluctuations. Thus, many researches believed the d-wave SC scenario at this stage.

However, the common belief was broken by recent experiments based on specific heat, thermal conductivity, and penetration depth measurements [21–24]. They revealed that fully gapped s-wave SC is realized in $CeCu_2Si_2$. In addition, T_c is quite robust against randomness. Therefore, s-wave SC state without sign reversal emerges in $CeCu_2Si_2$ [23]. These experimental results give a paradigm shift in the long history of research on superconductivity.

R. Tazai, *Theory of Multipole Fluctuation Mediated Superconductivity and Multipole Phase*, Springer Theses,
https://doi.org/10.1007/978-981-16-1026-4_4

In this chapter, we propose a microscopic theory of s-wave SC phase in $CeCu_2Si_2$. For this purpose, we have to solve the following primitive question:

- Why attractive SC pairing interaction overcome large Coulomb repulsion/AFM fluctuations in heavy fermion system?

4.2 Formalism

4.2.1 *Multi-Orbital Periodic Anderson Model*

Kinetic Term

We consider $4f^1$ ($L = 3, S = 1/2$) electrons on Ce-ion in $CeCu_2Si_2$. In the presence of strong SOI, $4f$ orbitals (14 folded) are split into $J = 3/2$ (8 folded) and $J = 5/2$ (6 folded) where J is total angular momentum. When we consider $4f^1$ (less than half) state, $J = 5/2$ becomes ground states. Furthermore, $J = 5/2$ states split into three Kramers doublets as shown in Fig. 4.1a. According to the LDA+DMFT study on $CeCu_2Si_2$[25], following two Kramers doublets have large DoS around the Fermi energy at ambient pressure. They are expressed in the J_z basis as

$$
\begin{aligned}
|f_1 \Downarrow\rangle &= a|+\frac{5}{2}\rangle + b|-\frac{3}{2}\rangle, \\
|f_1 \Uparrow\rangle &= a|-\frac{5}{2}\rangle + b|+\frac{3}{2}\rangle, \\
|f_2 \Uparrow\rangle &= -a|+\frac{3}{2}\rangle + b|-\frac{5}{2}\rangle, \\
|f_2 \Downarrow\rangle &= -a|-\frac{3}{2}\rangle + b|+\frac{5}{2}\rangle,
\end{aligned}
\tag{4.1}
$$

where $\Downarrow$ ($\Uparrow$) represents pseudo-spin up (down). a and $b(=\sqrt{1-a^2})$ are coefficient parameter determined by CEF. Here, the third Kramers doublet $|f_3\rangle = |J_z = \pm\frac{1}{2}\rangle$ is dropped since it gives negligibly small DoS around the Fermi level.

Then, we introduce Periodic Anderson Model (PAM). The kinetic term is given by

$$
\hat{H}_0 = \sum_{k\Sigma} \epsilon_k c^\dagger_{k\Sigma} c_{k\Sigma} + \sum_{kl\sigma} E_{lk} f^\dagger_{kl\sigma} f_{kl\sigma} + \sum_{kl\Sigma\sigma} \left(V^*_{kl\Sigma\sigma} f^\dagger_{kl\sigma} c_{k\Sigma} + V_{kl\Sigma\sigma} c^\dagger_{k\Sigma} f_{kl\sigma} \right), \tag{4.2}
$$

where $c^\dagger_{k\Sigma}$ ($c_{k\Sigma}$) is a creation (annihilation) operator for conductive electron with momentum $\boldsymbol{k}$, real spin Σ, and energy ϵ_k, which is given by

$$
\epsilon_k = 2t_{ss}(\cos k_x + \cos k_y) + \epsilon_0, \tag{4.3}
$$

where $t_{ss} = -1.0$. $f^{\dagger}_{kl\sigma}$ ($f_{kl\sigma}$) is a creation (annihilation) operator for f-electron with $\boldsymbol{k}$, orbital $l = 1, 2$, pseudo-spin σ, and energy $E_{l\boldsymbol{k}}$. $V_{\boldsymbol{k}l\Sigma\sigma}$ is the hybridization term between f-electron and s-electron. In this study, we consider 2D square lattice model as shown in Fig. 4.1b. Both f- and s-orbitals are on Ce-ion. $V_{\boldsymbol{k}l\Sigma\sigma}$ is calculated by using Slater-Koster table [45];

$$
\begin{aligned}
V_{\boldsymbol{k}f_1\uparrow\Uparrow} &= -\sqrt{\frac{3}{14}}t_{sf}(a\sqrt{5}+b)(\sin k_y - i\sin k_x),\\
V_{\boldsymbol{k}f_1\downarrow\Downarrow} &= \sqrt{\frac{3}{14}}t_{sf}(a\sqrt{5}+b)(\sin k_y + i\sin k_x),\\
V_{\boldsymbol{k}f_2\uparrow\Uparrow} &= \sqrt{\frac{3}{14}}t_{sf}(a-\sqrt{5}b)(\sin k_y - i\sin k_x),\\
V_{\boldsymbol{k}f_2\downarrow\Downarrow} &= -\sqrt{\frac{3}{14}}t_{sf}(a-\sqrt{5}b)(\sin k_y + i\sin k_x).
\end{aligned} \tag{4.4}
$$

Hereafter, to simplify the analysis, we put $a = 1, b = 0$. Then, we obtain

$$
V_{\boldsymbol{k}l\Sigma\sigma} = \sigma(-1)^l t^l_{sf}(\sin k_y - i\sigma\sin k_x)\delta_{\sigma,\Sigma}, \tag{4.5}
$$

where $\delta_{\sigma,\Sigma}$ is Kronecker delta function. Note that pseudo-spin is conserved ($V_{\uparrow\Downarrow} \neq 0$) since the present system is 2D and has space inversion symmetry. The detailed derivation of Eq. (4.5) is explained in Appendix 3. The imaginary terms come from strong SOI considered in the ground states in Eq. (4.1). We put $\mu = -5.52\times10^{-3}$, temperature $T = 0.02$, and the hopping parameter $t_{sf} = 0.7$. Then, f-electron number is $n_f = 0.9$ and s-electron number is $n_s = 0.3$. In this case,

$$
(t^{f_1}_{sf}, t^{f_2}_{sf}) = (0.724, 0.324), \tag{4.6}
$$

which means that the two orbitals have different itinerancy.

We show the obtained band structure, Fermi surface, and DoS in Fig. 4.2a, b and c, respectively. The chemical potential corresponds to $\epsilon = 0$. In addition, we set $E_{1\boldsymbol{k}} = 0.2$ and $E_{2\boldsymbol{k}} = 0.1$ by considering CEF splitting. $|t_{ss}|$ is of order 1eV since $W_D \sim 10$eV holds in $CeCu_2Si_2$ [28]. The width of quasi-particle band (=the lowest band) is $W^{qp}_D \sim 1$. Large Fermi surface is obtained and the relation $D^{f_1}(0) \simeq D^{f_2}(0)$ is satisfied.

Interaction Term

In addition, we introduce on-site Coulomb interaction among f-electrons:

$$
\hat{\mathrm{H}}_U = r\cdot\frac{1}{4}\sum_i\sum_{ll'mm'}\sum_{\sigma\sigma'\rho\rho'}U^{\sigma\sigma';\rho\rho'}_{ll';mm'}f^{\dagger}_{il\sigma}f_{il'\sigma'}f_{im\rho}f^{\dagger}_{im'\rho'}, \tag{4.7}
$$

where i is site index. $\hat{U}$ is the interaction matrix normalized as $U^{\Uparrow\Downarrow;\Uparrow\Downarrow}_{11;11} = 1$. Note that $\hat{U}$ in Eq. (4.7) is anti-symmetrized:

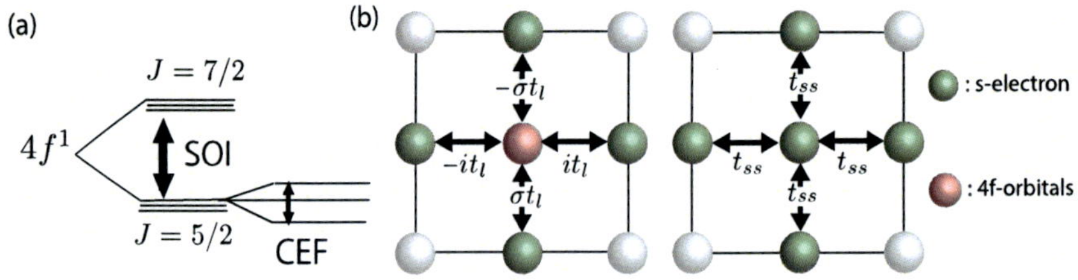

Fig. 4.1 **a** The nearest neighbor hopping integrals given by s-s and s-f hoppings. $\sigma = 1(-1)$ is pseudo-spin up (down) and $t_l \equiv (-1)^{l-1} t_{sf}^l$. Reprinted with permission from Ref. [55]. Copyright ©2018 by the American Physics Society

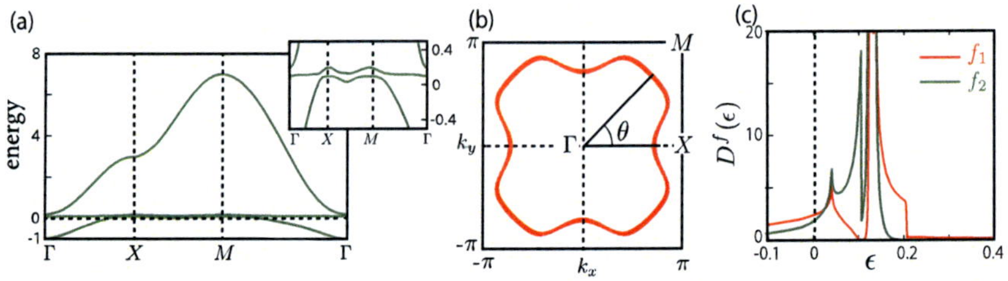

Fig. 4.2 **a** Band dispersion along high-symmetry line. **b** Obtained Fermi surface at $n_f = 0.9$. (c) Partial DoS of f_l-electrons. The red (green) line corresponds to $f_1(f_2)$-orbital. Reprinted with permission from Ref. [55]. Copyright ©2018 by the American Physics Society

$$U_{ll';mm'}^{\sigma\sigma';\rho\rho'} = (-1) U_{mm';ll'}^{\rho\rho';\sigma\sigma'}. \tag{4.8}$$

$\hat{U}$ is obtained by performing the unitary transformation of l_z-basis one, which is given by

$$\begin{aligned} \bar{U}_{l_z l_z', l_z'' l_z'''} &= \frac{e^2}{4\pi\epsilon_0} \int d\mathbf{r} \int d\mathbf{r}' \frac{u_{l_z}^*(\mathbf{r}) u_{l_z'''}^*(\mathbf{r}') u_{l_z'}(\mathbf{r}') u_{l_z''}(\mathbf{r})}{|\mathbf{r} - \mathbf{r}'|} \\ &= \sum_p a_{l_z l_z' l_z'' l_z'''}^p F^p, \end{aligned} \tag{4.9}$$

where $u_{l_z}(\mathbf{r})$ is the wave function of eigenstate of orbital angular momentum l_z.

$$u_{l_z}(\mathbf{r}) \equiv R(r)\Theta_{l_z}(\theta) e^{i l_z \phi}. \tag{4.10}$$

F^p is Slater integral introduced in Ref. [45], which is defined as

$$F^p = \frac{e^2}{4\pi\epsilon_0} \int dr \int dr' R^2(r) R^2(r') r_{\min}^p r_{\max}^{-(p+1)} r^2 r'^2, \tag{4.11}$$

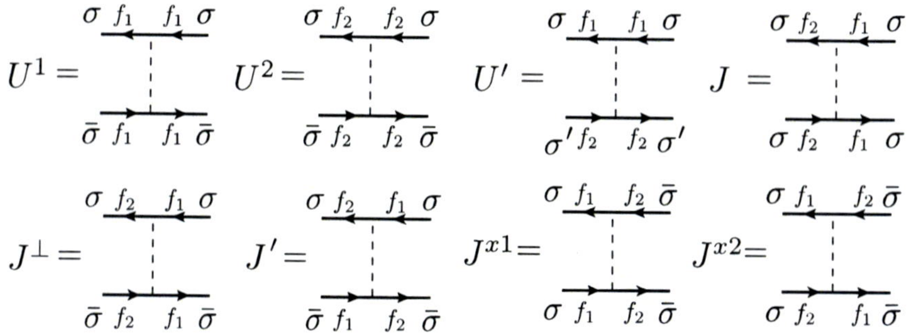

Fig. 4.3 Definition of multi-orbital Coulomb interaction in the pseudo-spin representation; $U^1, U^2, U', J, J^{\perp}, J', J^{x1}$, and J^{x2}

where $r_{\min} = \min\{r, r'\}$ and $r_{\max} = \max\{r, r'\}$. We put $(F^0, F^2, F^4, F^6) = (5.3, 9.09, 6.927, 4.756)$ in unit eV by referring Ref. [46]. The derivation of a^p is written in Appendix 3. Pseudo-spin is conserved in Eq. (4.7) while SU(2) is broken due to the SOI. In this case, $\hat{U}$ is uniquely decomposed into in-plane (s), out-of plane (s⊥) spin and charge (c) channel, which is defined as

$$\hat{U}^{\sigma\sigma';\lambda\lambda'} = \frac{1}{2}\hat{U}^{s}(\sigma^{x}_{\sigma\sigma'}\sigma^{x}_{\lambda'\lambda} + \sigma^{y}_{\sigma\sigma'}\sigma^{y}_{\lambda'\lambda}) + \frac{1}{2}\hat{U}^{s\perp}\sigma^{z}_{\sigma\sigma'}\sigma^{z}_{\lambda'\lambda} + \frac{1}{2}\hat{U}^{c}\sigma^{0}_{\sigma\sigma'}\sigma^{0}_{\lambda'\lambda}, \quad (4.12)$$

where $\boldsymbol{\sigma} = (\sigma^x, \sigma^y, \sigma^z)$ is Pauli matrix vector in the pseudo-spin space, and σ^0 is identity matrix. $\hat{U}^{s,s\perp,c}$ is defined as

$$\begin{cases} \hat{U}^{s} = \hat{U}^{\uparrow\uparrow;\uparrow\uparrow} - \hat{U}^{\uparrow\uparrow;\downarrow\downarrow} \\ \hat{U}^{s\perp} = \hat{U}^{\uparrow\downarrow;\uparrow\downarrow} \\ \hat{U}^{c} = \hat{U}^{\uparrow\uparrow;\uparrow\uparrow} + \hat{U}^{\uparrow\uparrow;\downarrow\downarrow}. \end{cases} \quad (4.13)$$

The matrix elements of $\hat{U}^{s,s\perp,c}$ are summarized in Table 4.1. They have general correspondence to SU(2) Hubbard model and composed of $\{U^{1(2)} U' J J^{\perp} J' J^{x1(2)}\}$ in Fig. 4.3.

- We introduce two-orbital PAM for describing $4f^1$ states.
- The s-f mixing term has imaginary component due to the SOI.
- Pseudo-spin is conserved in 2D model with space inversion symmetry.
- Coulomb interaction is expressed by s, s⊥, and c -channel.

Table 4.1 Matrix elements of Coulomb interaction for s, s⊥, and c-channel for $l \neq m$. $J = J'$, $J^{\perp} = 0$, and $J^{x1} = -J^{x2}$ are satisfied

s	Type	Value
$U^{0;s}_{11;11}$	U^1	1.0
$U^{0;s}_{22;22}$	U^2	0.90
$U^{0;s}_{lm;lm}$	$U' - J + J^{\perp}$	0.80
$U^{0;s}_{ll;mm}$	$J - J^{x1}$	−0.12
$U^{0;s}_{lm;ml}$	$J' - J^{x2}$	0.20
$s\perp$		
$U^{0;s\perp}_{11;11}$	U^1	1.0
$U^{0;s\perp}_{22;22}$	U^2	0.90
$U^{0;s\perp}_{lm;lm}$	$U' - J^{x1}$	0.68
$U^{0;s\perp}_{ll;mm}$	$J^{\perp}$	0.0
$U^{0;s\perp}_{lm;ml}$	$J' - J^{x2}$	0.20
c		
$U^{0;c}_{11;11}$	$-U^1$	−1.0
$U^{0;c}_{22;22}$	$-U^2$	−0.90
$U^{0;c}_{lm;lm}$	$U' - J - J^{\perp}$	0.80
$U^{0;c}_{ll;mm}$	$J - 2U' + J^{x1}$	−1.5
$U^{0;c}_{lm;ml}$	$-J' + J^{x2}$	−0.20

4.2.2 *Multipole Symmetry*

In the present model in Eq. (4.1), there are 16-type active multipole $Q(= 1 \sim 16)$. They are expressed as monopole (rank 0), dipole (rank 1), quadrupole (rank 2), octupole (rank 3), hexadecapole (rank 4), and dotriacontapole (rank 5) and classified into each irreducible representation (IR); ($\Gamma = A_1^+, A_2^+, E^+, A_1^-, A_2^-, E^-$) as shown in Table 4.2 [13]. Each multipole operator $\hat{O}^Q$ of rank k is composed of 4×4 tensor $J_q^{(k)}(q = -k \sim k)$[8, 47] which is defined as

$$[J_{\pm}, J_q^{(k)}] = \sqrt{(k \mp q)(k \pm q + 1)}J_{q\pm1}^{(k)} \tag{4.14}$$

$$J_k^{(k)} = (-1)^k \sqrt{\frac{(2k-1)!!}{(2k)!!}} J_+^k. \tag{4.15}$$

Detailed derivation of $\hat{O}^Q$ from $J_q^{(k)}$ is given in Appendix 3. The matrix representations of each $\hat{O}^Q$ are summarized in fourth column in Table 4.2. $\hat{\sigma}^{\mu}, \hat{\tau}^{\mu}(\mu = x, y, z)$

Table 4.2 Irreducible representation and 16-type active multipoles in the present two-orbital model. Operator with rank k corresponds to 2^k-pole. Each operator is classified into s, s⊥, and c-ch

IR (Γ)	Rank (k)	Multipole (Q)	Matrix ($\hat{O}^Q$)	ch
A_1^+	0	$\hat{1}$	$\hat{\sigma}^0\hat{\tau}^0$	c
	2	$\hat{O}_{20}$	$\hat{\sigma}^0(2.00\hat{\tau}^0+3.00\hat{\tau}^z)$	
	4	$\hat{H}_0$	$\hat{\sigma}^0(-5.73\hat{\tau}^0+11.5\hat{\tau}^z-12.8\hat{\tau}^x)$	
A_2^+	4	$\hat{H}_z$	$-19.8\hat{\sigma}^z\hat{\tau}^y$	s
E^+	2	$\hat{O}_{yz(zx)}$	$-(+)3.87\hat{\sigma}^{x(y)}\hat{\tau}^y$	$s_\perp$
A_1^-	5	$\hat{D}_4$	$29.8i\hat{\sigma}^0\hat{\tau}^y$	c
A_2^-	1	$\hat{J}_z$	$\hat{\sigma}^z(0.50\hat{\tau}^0+2.00\hat{\tau}^z)$	s
	3	$\hat{T}_z$	$\hat{\sigma}^z(9.00\hat{\tau}^0-1.50\hat{\tau}^z)$	
	5	$\hat{D}_z$	$-29.8\hat{\sigma}^z\hat{\tau}^x$	
E^-	1	$\hat{J}_{x(y)}$	$-1.12\hat{\sigma}^{x(y)}\hat{\tau}^x$	$s_\perp$
	3	$\hat{T}_{x(y)}$	$\hat{\sigma}^{x(y)}(3.75\hat{\tau}^0-3.75\hat{\tau}^z+5.03\hat{\tau}^x)$	
	5	$\hat{D}_{x(y)}$	$\hat{\sigma}^{x(y)}(23.0\hat{\tau}^0-6.56\hat{\tau}^z-3.14\hat{\tau}^x)$	

are Pauli matrix in the pseudo-spin and orbital basis (f_1, f_2), respectively. $\hat{\sigma}^0$, $\hat{\tau}^0$ are identity matrices. They are classified into TRS even or odd as follows,

$$\{\sigma^0\}, \{\tau^0, \tau^z, \tau^x\} = \text{TRS} \quad \text{even} \quad / \quad \{\sigma^z, \sigma^x, \sigma^y\}, \{\tau^y\}, = \text{TRS} \quad \text{odd}. \tag{4.16}$$

Note that the present two-orbital system has space inversion symmetry. Then, even (odd)-rank operators correspond to electric (magnetic) channel since they correspond to time reversal even (odd). The fifth column in Table 4.2 shows corresponding pseudo-spin channel of $\hat{O}^Q$, which is defined by Eq. (4.13).

Now, on-site Coulomb interaction in Eq. (4.7) is rewritten by using multipole:

$$U^{QQ'} = (\mathbf{O}^Q)^\dagger \hat{U} \mathbf{O}^{Q'}. \tag{4.17}$$

$\mathbf{O}^Q$ is 16×1 vector, where $(\mathbf{O}^Q)_{4(L-1)+M}$ denotes $(\hat{O}^Q)_{L,M}$. In Table 4.3, obtained value for U^{QQ} is shown. The magnetic Coulomb interaction $U^{QQ}(Q = J, T, D)$ is larger than electric ones $(Q = C, O, H)$. For this reason, RPA fails to derive electric phase, while higher order many body effects bring various electric phases.

- There are 16-type active multipole orders given by Q.
- Even (odd)-rank Q is corresponding to TRS even (odd).
- Magnetic ch Coulomb interactions are larger than electric ones.
- Higher rank Coulomb interactions are as large as dipole ones.

Table 4.3 Normalized Coulomb interaction $U^Q (\equiv U^{QQ})$ for multipole channel Q. Reprinted with permission from Ref. [55]. Copyright ©2018 by the American Physics Society

Q	C	O_{20}	H_0	H_z	$O_{yz(zx)}$	
U_0^Q	-1.3	-0.18	0.17	0.34	0.27	
Q	J_z	T_z	$D_{z(4)}$	$J_{x(y)}$	$T_{x(y)}$	$D_{x(y)}$
U_0^Q	0.56	0.44	0.55	0.49	0.49	0.50

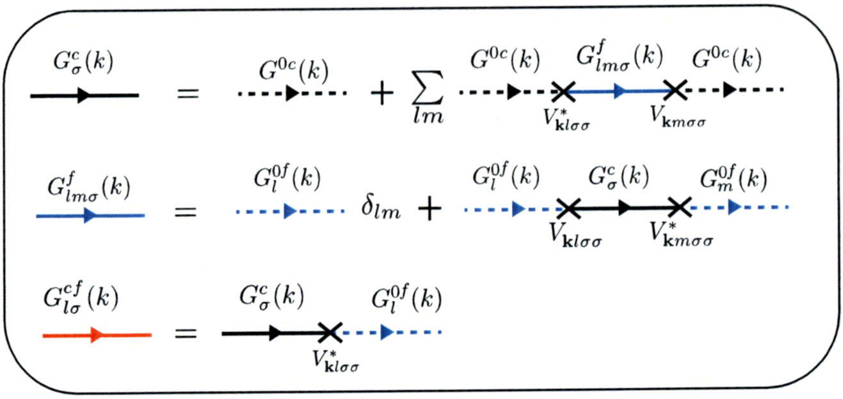

Fig. 4.4 Dyson equation of the present model

Green Function

Here, we introduce one-particle Green functions defined by Hamiltonian in Eq. (4.2). The Green functions without s-f mixing are given as

$$G^{0c}(k) = \frac{1}{i\epsilon_n - \epsilon_k + \mu} \tag{4.18}$$

$$G^{0f}_l(k) = \frac{1}{i\epsilon_n - E_{lk} + \mu}. \tag{4.19}$$

By solving Dyson equation (diagrammatic expression is shown in Fig. 4.4), we obtain the following Green functions including s-f mixing:

$$G^c_\sigma(k) = G^{0c}(k) + G^{0c}(k)\sum_{lm}(V^*_{kl\sigma\sigma}G^f_{lm\sigma}(k)V_{km\sigma\sigma})G^{0c}(k), \tag{4.20}$$

$$G^f_{lm\sigma}(k) = G^{0f}_l(k)\delta_{lm} + G^{0f}_l(k)V_{kl\sigma\sigma}G^c_\sigma(k)V^*_{km\sigma\sigma}G^{0f}_m(k), \tag{4.21}$$

$$G^{cf}_{l\sigma}(k) = G^c_\sigma(k)V^*_{kl\sigma\sigma}G^{0f}_l(k), \tag{4.22}$$

where (l, m) takes 1 or 2, $k = (\boldsymbol{k}, \epsilon_n) = (\boldsymbol{k}, (2n+1)\pi T)$ and μ is the chemical potential. In the present two-orbital model, we neglect the self-energy due to the Coulomb interaction among f-electrons. It works as renormalization factor for the energy scale of the electron system. $V^*_{kl\sigma\sigma}V_{km\sigma\sigma}$ in Eqs. (4.20) and (4.21) are given by

$$V^*_{kl\sigma\sigma} V_{km\sigma\sigma} = t^l_{sf} t^m_{sf} (-1)^{l+m} (\sin^2 k_y + \sin^2 k_x). \tag{4.23}$$

Then, we obtain

$$V^*_{kl\uparrow\uparrow} V_{km\uparrow\uparrow} = V^*_{kl\downarrow\downarrow} V_{km\downarrow\downarrow}. \tag{4.24}$$

For this reason, the Green functions G^c and G^f are independent of spin index:

$$\begin{aligned} G^f_{lm}(k) &\equiv G^f_{lm\uparrow}(k) = G^f_{lm\downarrow}(k), \\ G^c(k) &\equiv G^c_\uparrow(k) = G^c_\downarrow(k). \end{aligned} \tag{4.25}$$

We comment that $\hat{G}^f(k)$ is written by using $\hat{\sigma}$ and $\hat{\tau}$, which is described as

$$\hat{G}^f \propto \hat{\sigma}^0(\hat{\tau}^0 - c\hat{\tau}^x), \tag{4.26}$$

where c denotes the coefficient parameter. Therefore, $\hat{G}^f(k)$ is classified into electric ch based on Eq. (4.17).

- Green functions with s-f mixing $\hat{G}^f$, $\hat{G}^c$, $\hat{G}^{cf}$ are given by Dyson equation.
- $\hat{G}^f$ and $\hat{G}^c$ are independent of spin index.
- $\hat{G}^f$ is proportional to $\hat{\sigma}^0(\hat{\tau}^0 - c\hat{\tau}^x)$.

4.2.3 *Multipole Susceptibility*

In this section, we calculate f-electrons susceptibility. First, the irreducible susceptibility (loop diagram) is defined as

$$\chi^0_{ll'mm'}(q) = \int_0^\beta d\tau \left\langle T_\tau \, f^\dagger_{kl'\sigma}(\tau) f_{k+ql\sigma}(\tau) f^\dagger_{k+qm\sigma}(0) f_{km'\sigma}(0) \right\rangle_{\hat{H}_0}, \tag{4.27}$$

where $\omega_j = 2j\pi T$ is the Boson Matsubara frequency and T_τ is T-product for fermion, where

$$\begin{aligned} T_\tau A_{\tau_1} A_{\tau_2} &= A_{\tau_1} A_{\tau_2} \;\; (\tau_1 > \tau_2), \\ &= -A_{\tau_2} A_{\tau_1} \;\; (\tau_1 < \tau_2). \end{aligned} \tag{4.28}$$

By using the Green function, it is rewritten as

$$\chi^0_{ll'mm'}(q) = -T\sum_k G^f_{lm}(k+q)G^f_{m'l'}(k), \tag{4.29}$$

where $q = (\boldsymbol{q}, \omega_j) = (\boldsymbol{q}, 2j\pi T)$. Note that $\hat{\chi}^0(q)$ is independent of spin index due to Eq. (4.19). Next, we define total f-electrons susceptibility considering the Coulomb interaction in Table 4.1.

$$\chi^{\sigma\sigma'\gamma\gamma'}_{ll'mm'}(q) = \int_0^\beta d\tau \left\langle T_\tau\ f^\dagger_{\boldsymbol{k}l'\sigma}(\tau) f_{\boldsymbol{k}+\boldsymbol{q}l\sigma}(\tau) f^\dagger_{\boldsymbol{k}+\boldsymbol{q}m\sigma}(0) f_{\boldsymbol{k}m'\sigma}(0)\right\rangle_{\hat{H}_0+\hat{H}_U}. \tag{4.30}$$

If we perform the RPA, susceptibilities in pseudo-spin channel are given by

$$\hat{\chi}^{ch}(q) = \hat{\chi}^0(q)(\hat{1} - u\hat{U}^{ch}\hat{\chi}^0(q))^{-1}, \tag{4.31}$$

where $ch = s, s\perp, c$. $\hat{\chi}^0(q)$, $\hat{\chi}^{ch}(q)$, and $\hat{U}^{ch}$ are 4×4 matrices. The definition of $\hat{U}^{ch}$ is given in Eq. (4.13). Here, we define the pseudo-spin Stoner factor α^{ch}_S as the largest eigenvalue of $u\hat{U}^{ch}\hat{\chi}^0(q)$, which is defined as

$$\lambda^i_{ch,q}\mathbf{x}^i_{ch,q} = u\hat{U}^{ch}\hat{\chi}^0(q)\mathbf{x}^i_{ch,q}, \tag{4.32}$$

$$\alpha_S \equiv \max_{ch}\{\alpha_{ch}\} \equiv \max_{q,i}\left\{\lambda^i_{ch,q}\right\}, \tag{4.33}$$

where $\lambda^i(\mathbf{x}^i)$ denotes the ith eigenvalue (eigenvector) of Eq. (4.32) In the present model, each matrix elements of $\hat{U}^s$ and $\hat{U}^{s\perp}$ in Table 4.1 are almost same except for $(lmlm)$ and $(llmm)$ elements. For this reason, $\hat{\chi}^s \approx \hat{\chi}^{s\perp}$ and $\alpha^s_S \approx \alpha^{s\perp}_S$ are satisfied.

- Susceptibility $\hat{\chi}^{ch}$ and Stoner factor α^{ch}_S are defined in $s, s\perp, c$ basis.

Multipole Susceptibility
Now, we define the multipole susceptibility for the channel Q:

$$\chi^{Q,Q'}(q) \equiv \sum_{\substack{\boldsymbol{k}ll'mm'\\ \sigma\sigma'\gamma\gamma'}} \int_0^\beta d\tau e^{i\omega_j\tau} \times \left\langle T_\tau\ f^\dagger_{\boldsymbol{k}l'\sigma'}(\tau) O^Q_{l\sigma;l'\sigma'} f_{\boldsymbol{k}+\boldsymbol{q}l\sigma}(\tau)\ f^\dagger_{\boldsymbol{k}+\boldsymbol{q}m\gamma}(0) O^{Q'}_{m'\gamma';m\gamma} f_{\boldsymbol{k}m'\gamma'}(0)\right\rangle, \tag{4.34}$$

where $\sigma, \sigma'\gamma, \gamma'$ are index of pseudo-spin. In the present 2D model,

$$\chi^{Q,Q'}(q) = 0 \quad \text{for} \quad \Gamma \neq \Gamma' \quad (Q\in\Gamma,\ Q'\in\Gamma'). \tag{4.35}$$

On the other hand, in 3D models, $\chi^{Q,Q'}(q) \neq 0$ even if $\Gamma \neq \Gamma'$. $\chi^{Q,Q'}$ is classified into magnetic (TRS odd) or electric (TRS even) channel, which is expressed as

$$\begin{aligned} &\text{Magnetic ch;} \quad \chi^{Q,Q'}(q) \quad \text{for} \quad Q, Q' \in A_1^+, A_2^+, E^+ \\ &\text{Electric ch;} \quad \chi^{Q,Q'}(q) \quad \text{for} \quad Q, Q' \in A_1^-, A_2^-, E^-. \end{aligned} \tag{4.36}$$

Note that the the absolute value of $\chi^{Q,Q'}(q)$ becomes larger as rank of Q gets higher. To compare the different rank multipole susceptibilities, we define normalized operator as

$$\hat{\Theta}^Q \equiv \hat{O}^Q / \sqrt{\mathrm{Tr}(\hat{O}^Q \hat{O}^{Q*})}. \tag{4.37}$$

Then, the normalized susceptibility is given as

$$\chi_{\Theta}^{Q,Q'}(q) \equiv \quad \chi^{Q,Q'}(q) \quad (\hat{O}^Q, \hat{O}^{Q'} \text{in Eq. (4.35)} \rightarrow \hat{\Theta}^Q, \hat{\Theta}^{Q'}). \tag{4.38}$$

In addition, we define 16×16 multipole susceptibility $\hat{\chi}^{Q,Q'}(q)$ with orbital and pseudo-spin index,

$$\begin{aligned} &\chi_{ll'mm'}^{Q,Q'\sigma\sigma'\gamma\gamma'}(q) \equiv a^{Q,Q'}(q) \;\; O_{l\sigma;l'\sigma'}^{Q} O_{m'\gamma;m\gamma'}^{Q'}, \\ \Leftrightarrow \quad &\hat{\chi}^{Q,Q'}(q) \equiv a^{Q,Q'}(q) \, \mathbf{O}^Q \, \mathbf{O}^{Q'\dagger}, \end{aligned} \tag{4.39}$$

where $a^{Q,Q'}(q)$ is calculated as follows. First, we solve the following characteristic equation:

$$\hat{\chi}(q)\mathbf{v}^i(q) = \lambda^i(q)\mathbf{v}^i(q), \tag{4.40}$$

where $\lambda^i(q)$ is ith real eigenvalue ($i = 1 \sim 16$) and $\mathbf{v}^i(q)$ is 16-dimensional eigenvector normalized as $|\mathbf{v}^i| = 1$. In the present model, $\mathbf{v}^i(q)$ is uniquely characterized by IR. Then, $\mathbf{v}^i(q)$ for $i \in \Gamma$ is expanded in the basis of the multipole matrices as follows:

$$\mathbf{v}^i(q) = \sum_{Q \in \Gamma} b^{i,Q}(q)\mathbf{O}^Q \qquad i \in \Gamma, \tag{4.41}$$

where the coefficient $b^{i,Q}(q)$ is uniquely determined. Note that $\{\mathbf{O}^Q\}$ forms complete basis but not orthogonal basis within the same Γ. Then, $a^{Q,Q'}$ is given by

$$a^{Q,Q'}(q) = \sum_{i \in \Gamma} b^{i,Q}(q)\lambda^i(q)b^{i,Q'*}(q). \tag{4.42}$$

As a result, $\hat{\chi}^{Q,Q'}(q)$ in Eq. (4.39) is obtained. $\hat{\chi}(q)$ is reproduced as

$$\hat{\chi}(q) = \sum_{Q,Q'} \hat{\chi}^{Q,Q'}(q). \tag{4.43}$$

Note that $\hat{\chi}^{Q,Q'}(q)$ is independent of normalized condition of the multipole operator $\hat{O}^Q$ since it reproduces 16×16 matrix $\hat{\chi}(q)$, which is a great merit in analysis.

- Three multipole susceptibilities $\chi^{Q,Q'}$, $\chi_{\Theta}^{Q,Q'}$, and $\hat{\chi}^{Q,Q'}$ are introduced.
- In the case of $\Gamma \neq \Gamma'$, $\chi^{Q,Q'}$, $\chi_{\Theta}^{Q,Q'}$, $\hat{\chi}^{Q,Q'}$ go to zero for $Q \in \Gamma$, $Q' \in \Gamma'$.
- $\chi_{\Theta}^{Q,Q'}$ is used for comparing different rank of multipole susceptibilities.
- $\hat{\chi}^{Q,Q'}$ is determined irrespective of normalization condition of $\mathbf{O}^Q$.

Susceptibility by RPA
Here, we perform RPA based on multi-orbital PAM. In this calculation, we use 32×32 $\boldsymbol{k}$-meshes and 128 Matsubara frequencies. In Fig. 4.5a, we show $\boldsymbol{q}$-dependence of $\chi^{Q,Q'}(\boldsymbol{q}, 0)$ at $u = 0.31$ ($\alpha_S = 0.90$) for the magnetic dipole ch as Q, Q'; $\chi^{J_z,J_z}(\boldsymbol{q}, 0) \in A_2^-$ and $\chi^{J_x,J_x}(\boldsymbol{q}, 0) \in E^-$. We find that $\chi^{J_z,J_z}(\boldsymbol{q}, 0)$ is much larger than $\chi^{J_x,J_x}(\boldsymbol{q}, 0)$ at $\boldsymbol{q} = (0, 0)$ while they are almost the same around the peak at $\boldsymbol{q} \simeq (\pi/2, \pi/2)$. Thus, the uniform magnetic susceptibility shows strong Ising anisotropy, which is actually observed in $CeCu_2Si_2$.

$$\begin{aligned} \chi^{J_z,J_z}(\boldsymbol{q}, 0) &\gg \chi^{J_x,J_x}(\boldsymbol{q}, 0) \ \ \text{at} \boldsymbol{q} = (0, 0) \\ \chi^{J_z,J_z}(\boldsymbol{q}, 0) &\simeq \chi^{J_x,J_x}(\boldsymbol{q}, 0) \ \ \text{at} \boldsymbol{q} \simeq (\pi/2, \pi/2). \end{aligned} \tag{4.44}$$

We verified that the peak at $\boldsymbol{q} \simeq (\pi/2, \pi/2)$ originates from nesting vector on the large Fermi surface as shown in Fig. 4.2. In this result, α_S is given by magnetic (=odd-rank) susceptibility $\chi^{Q,Q'}(\boldsymbol{q}, 0)$ for $Q, Q' \in A_2^-$, that is, $\alpha_S = \alpha_{A_2^-}$. Then,

$$1 \gtrsim \alpha_{A_2^-} \gtrsim \alpha_{E^-}. \tag{4.45}$$

Next, we calculate $\chi_{\Theta}^{Q,Q'}(q)$ to compare among the different ranks of multipole susceptibility. In Fig. 4.5b, we show α_S dependences of the maximum value of magnetic multipole susceptibilities $\chi_{\max}^Q \equiv \max_{\boldsymbol{q}}\{\chi_{\Theta}^{Q,Q}(\boldsymbol{q}, 0)\}$. α_S linearly increases in proportional to u. The most divergent $\chi_{\max}^Q$ is that for $Q = T_x$. This fact is consistent with RPA result based on the first-principles model in Ref[28]. Secondly, $\chi_{\max}^Q$ for $Q = D_z, J_x, T_z, D_4$ is also strongly enlarged. Therefore, various magnetic multipole (including higher rank) susceptibilities are simultaneously enlarged in RPA. This is a characteristic feature of f-electron systems with strong SOI [13]. We find that the off-diagonal magnetic multipole susceptibilities, such as $\chi_{\Theta}^{J_z,T_z}$, are also enlarged.

- We obtain Ising-like dipole susceptibility $\chi^{J_z,J_z}(\boldsymbol{q}, 0) \gg \chi^{J_x,J_x}(\boldsymbol{q}, 0)$ at $\boldsymbol{q} = (0, 0)$, while they are almost the same at $\boldsymbol{q} \simeq (\pi/2, \pi/2)$.
- Higher rank magnetic multipole fluctuations develop by strong SOI.

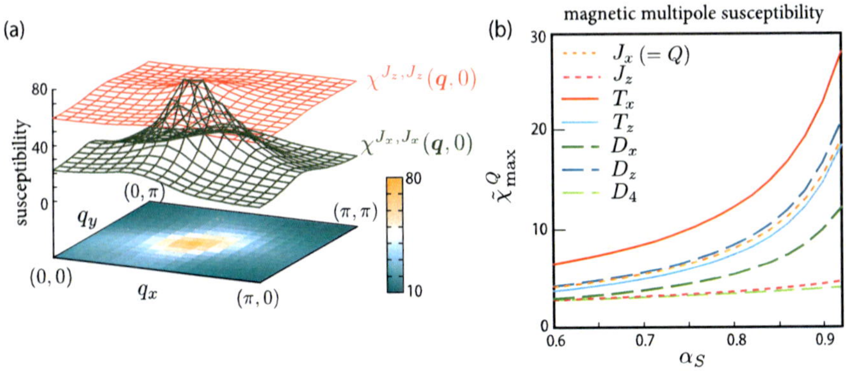

Fig. 4.5 **a** q dependence of the magnetic dipole susceptibility. $\chi^{J_z,J_z}(q,0) \gg \chi^{J_x,J_x}(q,0)$ is satisfied at $q = (0,0)$. **b** α_S dependence of magnetic multipole susceptibility. Higher rank magnetic multipole susceptibilities are strongly enlarged as well as dipole ones. Reprinted with permission from Ref. [55]. Copyright ©2018 by the American Physics Society

4.2.4 Phonon-Mediated Interaction

In this section, we introduce A_1^+ (= identical representation) phonon-mediated interaction. The effective interaction is introduced by

$$\hat{V}^{\rm ph} = 2g(\omega_j)\mathbf{C}_{A_1^+}\mathbf{C}^\dagger_{A_1^+}, \tag{4.46}$$

$$g(\omega_j) \equiv \tilde{g}\frac{\omega_D^2}{\omega_D^2+\omega_j^2},$$

where $\hat{V}^{\rm ph}$ is 16×16 matrix and $\tilde{g} = \frac{2\eta^2}{\omega_D}$ (>0). ω_D is the phonon frequency. η is the coupling constant between electrons and phonon. $\hat{C}_{A_1^+}$ is the 16×1 tensor where $(\mathbf{C}_{A_1^+})_{4(L-1)+M} \equiv (\hat{C}_{A_1^+})_{L,M}$ for $L=(l,\sigma)$. $\hat{C}_{A_1^+}$ is given by a linear combination of multipole operators belonging to A_1^+ in Table 4.2. It is expressed as

$$\hat{C}^{A_1^+} \equiv \hat{\sigma}^0(\alpha\hat{\tau}^0 + \beta\hat{\tau}^z + \gamma\hat{\tau}^x), \tag{4.47}$$

which includes monopole $\hat{1}$, quadrupole $\hat{O}_{20}$, and hexadecapole $\hat{H}_0$. For instance, the A_1^+ interaction can be induced by oscillation of c-axis length [48]. The schematic picture of the A_1^+ oscillation is shown in Fig. 4.6. In this case, the bare Coulomb interaction $\hat{U}$ is replaced with $\hat{U}^{eff}$:

$$u\hat{U} \to u\hat{U} + 2g(0)\mathbf{C}_{A_1^+}\mathbf{C}^\dagger_{A_1^+}. \tag{4.48}$$

This replacement enhances electric (=even-rank) multipole fluctuations since the A_1^+ interaction is classified into even-rank multipole interaction. On the other hand,

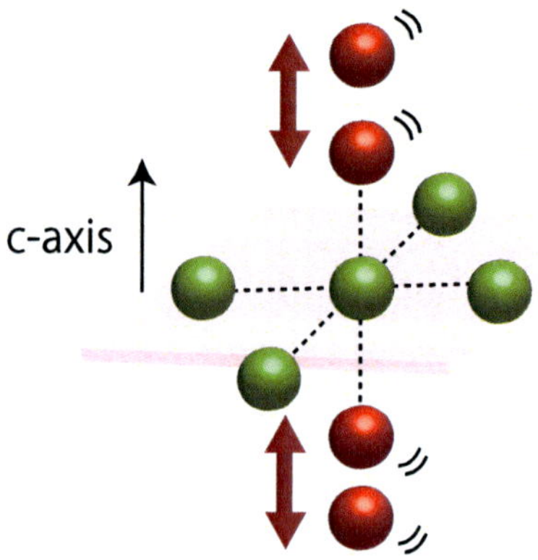

Fig. 4.6 **a** Schematic picture of A_1^+ phonon-mediated interaction. Reprinted with permission from Ref. [55]. Copyright ©2018 by the American Physics Society

the magnetic (=odd-rank) multipole susceptibilities are independent of the phonon-mediated interaction.

Hereafter, we mainly show the numerical result of $(\alpha, \beta, \gamma) = (0, 1, -1)$. We verified that the main results are qualitatively same as those of $(\alpha, \beta, \gamma) = (0, 1, 1)$. In the present study, we drop ω_j-dependence of $g(\omega_j)$ for simplicity, that is, we neglect the retardation effect. This fact leads to underestimation of the attractive superconducting pairing interaction or T_c of s-wave SC phase. Note that T_c of s-wave SC will increase, if the retardation effect is considered.

- We consider A_1^+ phonon-mediated interaction.
- A_1^+ interaction induces electric $(\hat{1}, \hat{O}_{20}, \hat{H}_0)$ multipole fluctuations.
- T_c will be enhanced if the retardation effect is considered.

4.3 Results

4.3.1 SC Phase Diagram

In this section, we solve the SC gap equation within Migdal-Eliashberg approximation. First, we define SC pairing interaction $\hat{V}^{ch}$ in $s, s\perp, c$ basis, which is given as

$$\hat{V}^{ch}(q) \equiv u^2 \hat{U}^{ch} \hat{\chi}^{ch}(q) \hat{U}^{ch} + u \hat{U}^{ch}. \quad (4.49)$$

Then, spin singlet pairing interaction $\hat{V}^{\mathrm{sing}}$ is given by

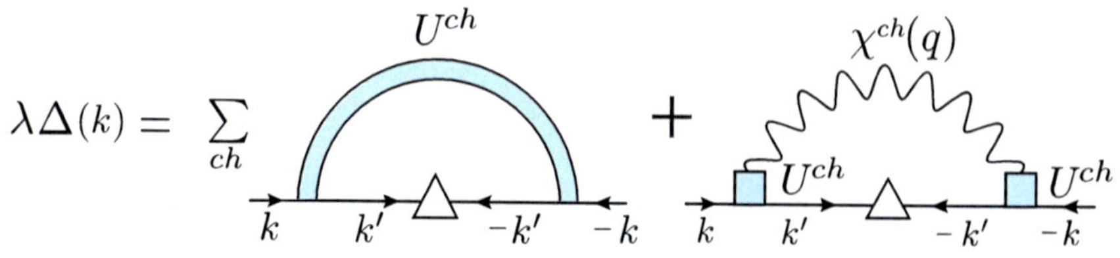

Fig. 4.7 The linearized gap equation within Migdal-Eliashberg approximation. Blue rectangles are bare Coulomb interaction U

$$\hat{V}^{\text{sing}}(q) = \hat{V}^{\text{sL}}(q) + \frac{1}{2}\hat{V}^{s}(q) - \frac{1}{2}\hat{V}^{c}(q) - u\hat{U}^{\text{sL}}. \tag{4.50}$$

The fourth term is needed to avoid double counting. Therefore, the linearized gap equation in the orbital basis is given by

$$\lambda\Delta_{ll'}(k) = -T \sum_{k',mm'nn'} G^{f}_{mn}(k')G^{f}_{m'n'}(-k')\Delta_{nn'}(k')V^{\text{sing}}_{lmm'l'}(k-k'), \tag{4.51}$$

where $\Delta_{ll'}(k)$ is the gap function and λ is the eigenvalue. Note that $\lambda = 1$ at $T = T_c$. The diagrammatic expression of the gap equation is shown in Fig. 4.7. In Eq. (4.51), V^{s} and V^{sL} bring sign change between $\Delta(k)$ and $\Delta(k')$, while uniform (s-wave) gap function is induced by V^{c}.

Next, we derive the gap equation in the band basis, which is given by

$$\lambda\Delta_{a}(k) = -T \sum_{k',b} G_{b}(k')G_{b}(-k')\Delta_{b}(k')V^{\text{sing}}_{ab}(k,k'), \tag{4.52}$$

where a, b is band index. After the integration along energy axis, we obtain

$$\lambda\Delta(\boldsymbol{k}, \epsilon_n) = -\frac{\pi T}{(2\pi)^2}\sum_{\epsilon_m}\oint \frac{d\boldsymbol{k}'}{v_{\boldsymbol{k}'}}\frac{\Delta(\boldsymbol{k}', \epsilon_m)}{|\epsilon_m|}V^{\text{sing}}(k,k'), \tag{4.53}$$

where $\Delta(\boldsymbol{k}, \epsilon_n)$ is the gap function on Fermi surface and $v_{\boldsymbol{k}}$ is Fermi velocity. Here, the band index is omitted since we consider one-band system as shown in Fig. 4.2. The band basis pairing interaction in Eq. (4.53) is obtained as follows:

$$V^{\text{sing}}(k,k') \equiv V^{\uparrow\downarrow\uparrow\downarrow}(k,k') - V^{\uparrow\uparrow\downarrow\downarrow}(k,k'), \tag{4.54}$$

where $\uparrow$ or $\downarrow$ denote pseudo-spin of the Kramers doublet in the Bloch function. $V^{\uparrow\downarrow\uparrow\downarrow(\uparrow\uparrow\downarrow\downarrow)}$ is obtained by performing the unitary transformation of the pairing interaction in the orbital basis

$$V^{\uparrow\downarrow\uparrow\downarrow}_{(\uparrow\uparrow\downarrow\downarrow)}(k,k') = \sum_{\substack{ll'mm' \\ \sigma\sigma'\gamma\gamma'}} V^{\sigma\sigma'\gamma\gamma'}_{ll'mm'}(k,k') u^{\uparrow *}_{l\sigma}(\boldsymbol{k}) u^{\downarrow *}_{m'\lambda'}(-\boldsymbol{k}) u^{\uparrow(\downarrow)}_{m\lambda}(-\boldsymbol{k}') u^{\downarrow(\uparrow)}_{l'\sigma'}(\boldsymbol{k}') \tag{4.55}$$

where $u^{\uparrow(\downarrow)}_{l\sigma}(\boldsymbol{k})$ is the unitary matrix connecting between $f^{\dagger}_{\boldsymbol{k}l\sigma}$ and the quasi-particle creation operator $f^{\dagger}_{\boldsymbol{k}\uparrow(\downarrow)}$. In the presence of the time reversal symmetry, following relation is satisfied [49]:

$$u^{\uparrow}_{l\sigma}(-\boldsymbol{k}) = (2\delta_{\uparrow\sigma} - 1) u^{\downarrow}_{l\bar{\sigma}}(\boldsymbol{k})^{*}. \tag{4.56}$$

Essentially, s and $s\perp$-channel interactions work as repulsion, while c-channel one works as attraction since the following equation is derived by neglecting the SOI:

$$V^{\mathrm{sing}}(k,k') = \left\{ \hat{V}^{s\perp}(k,k') + \frac{\hat{V}^{s}(k,k')}{2} - \frac{\hat{V}^{c}(k,k')}{2} \right\} \bar{u}^{*}_{l}(\boldsymbol{k}) \bar{u}_{m'}(\boldsymbol{k}) \bar{u}^{*}_{m}(\boldsymbol{k}') \bar{u}_{l'}(\boldsymbol{k}').$$

Here, $u^{\uparrow(\downarrow)}_{l\sigma}(\boldsymbol{k}) = \delta_{\uparrow(\downarrow),\sigma} \bar{u}_{l}(\boldsymbol{k})$ is assumed.

Finally, we define the multipole-decomposed SC pairing interaction in the same way as $\hat{\chi}^{Q,Q'}$ in Eq. (4.39). It is expressed as

$$\hat{V}^{ch,QQ'}(q) \equiv d^{ch,Q,Q'}(q) \mathbf{O}^{Q} \mathbf{O}^{Q'\dagger} \quad \left(\hat{V}^{QQ'}(q) \equiv d'^{Q,Q'}(q) \mathbf{O}^{Q} \mathbf{O}^{Q'\dagger} \right), \tag{4.57}$$

where $d^{Q,Q'}$ and $d'^{Q,Q'}$ are obtained by solving the characteristic equation:

$$\hat{V}^{ch}(q) \mathbf{v}^{i,ch}(q) = \lambda^{i,ch}(q) \mathbf{v}^{i,ch}(q), \quad \left(\hat{V}(q) \mathbf{v}^{i}(q) = \lambda^{i}(q) \mathbf{v}^{i}(q) \right). \tag{4.58}$$

Note that $\hat{V}^{ch,QQ'} = 0$ for $ch \notin \Gamma$ in the case of $Q, Q' \in \Gamma$.

- We define linearized gap equation within Migdal-Eliashberg approximation.
- s, $s\perp$-ch (c-ch) interaction works as repulsive (attractive) pairing interaction.
- SC pairing interaction is decomposed into multipole ch.

4.3.2 U-VC

In this section, we consider higher order many body effects beyond Migdal-Eliashberg approximation. They are expressed as the three-point vertex correction for bare Coulomb interaction U. Therefore, we call the vertex corrections U-VC. In the present model, U-VC satisfies the pseudo-spin conservation. Thus, the pairing interaction for s, $s\perp$, c-channel in Eq. (4.50) is replaced with

$$\hat{V}^{ch}_{U-VC}(k,k') = \hat{\Lambda}^{ch}_{k,k'}\hat{V}^{ch}(k-k')\hat{\Lambda}^{ch\dagger}_{-k,-k'}, \tag{4.59}$$

$$(\hat{\Lambda}^{ch\dagger}_{k,k'})_{ll'mm'} \equiv (\hat{\Lambda}^{ch}_{k,k'})_{m'ml'l}. \tag{4.60}$$

Here, $\hat{V}^{ch}(q)$ is given in Eq. (4.49). $\hat{\Lambda}^{ch}_{k,k'}$ is the enhancement factor due to AL-type U-VC given by $\hat{L}^{ch}_{k,k'}$, whose diagrammatic expression is shown in Fig. 4.8. In the present study, MT-type U-VC is negligible compared to AL-type one. For this reason, we calculate only AL-type U-VC.

$$(\hat{\Lambda}^{ch}_{k,k'})_{ll'mm'} = \delta_{lm}\delta_{l'm'} + (\hat{L}^{ch}_{k,k'})_{ll'mm'}. \tag{4.61}$$

The second term is neglected within Migdal-Eliashberg approximation. Analytic expression of $\hat{L}^{ch}_{k,k'}$ is given as

$$\begin{aligned}(\hat{L}^{ch}_{k,k'})_{ll'mm'} &\equiv \frac{T}{2}\sum_{\substack{p,abcdef\\ ch1,ch2}} B^{mm'}_{abcdef}(k-k',p)a^{ch}_{ch1,ch2}V^{ch1}_{lacd}(k-k'+p)V^{ch2}_{bl'ef}(-p),\\ \Leftrightarrow \quad \hat{L}^{ch}_{k,k'} &= \frac{T}{2}\sum_{p,ch1,ch2} a^{ch}_{ch1,ch2}\hat{B}(k-k',p)\hat{V}^{ch1}(k-k'+p)\hat{V}^{ch2}(-p), \end{aligned} \tag{4.62}$$

where

$$\begin{cases} B^{mm'}_{abcdef}(q,p) = \frac{1}{4}G^{f}_{ab}(k'-p)\left\{C^{mm'}_{cdef}(q,p) + C^{mm'}_{efcd}(q,q+p)\right\}, \\ C^{ab}_{cdef}(q,p) = -T\sum_{k'} G^{f}_{ca}(k'+q)G^{f}_{bf}(k')G^{f}_{ed}(k'-p). \end{cases} \tag{4.63}$$

Here, $a \sim f$ are orbital indices while p denotes wave number. $a^{ch}_{ch1,ch2}$ is coefficient parameter at

$$(a^{c}_{s,s}, a^{c}_{s\perp,s\perp}, a^{c}_{c,c}) = (1,2,1). \tag{4.64}$$

Then, we obtain

$$\hat{L}^{c} \propto \hat{V}^{s}\hat{V}^{s} + 2\hat{V}^{s\perp}\hat{V}^{s\perp} + \hat{V}^{c}\hat{V}^{c}. \tag{4.65}$$

We comment that $\hat{V}^{s}\hat{V}^{s} + 2\hat{V}^{s\perp}\hat{V}^{s\perp}$ is enhanced by magnetic QCP due to the development of spin fluctuations. In conclusion, charge-ch of U-VC is enlarged by the square of spin fluctuations $O(\chi^2)$ near the magnetic QCP. We call this nontrivial effect "mode-coupling" between spin and charge channel, which originates from many body effects beyond Migdal-Eliashberg approximation.

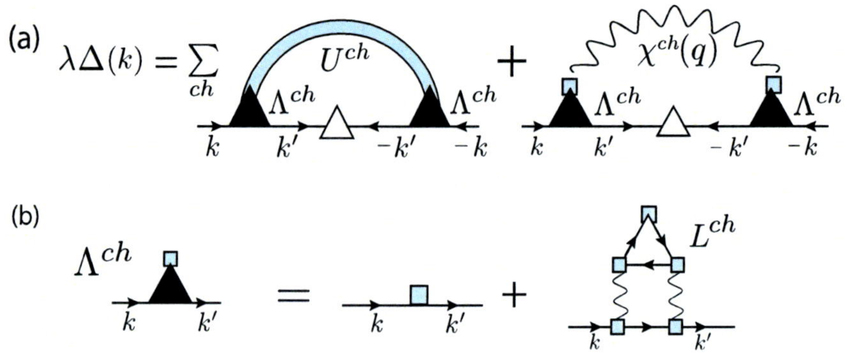

Fig. 4.8 **a** Linearized gap equation in the present study. The black triangle is three-point vertex correction (U-VC). **b** U-VC due to the AL process

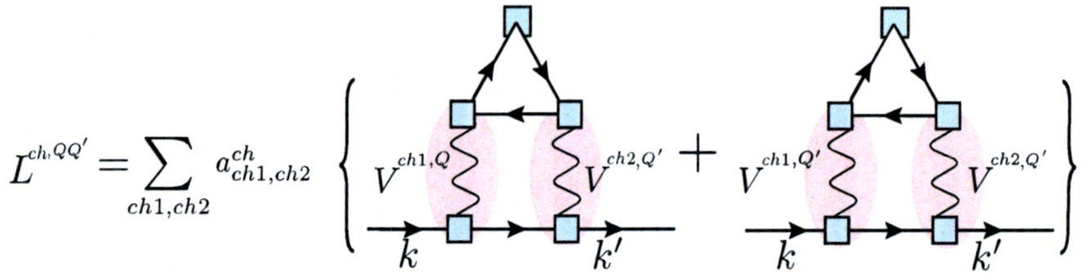

Fig. 4.9 Multipole-decomposed U-VC for Q, Q'

Next, we define multipole-decomposed U-VC, which is expressed as

$$(\hat{\Lambda}_{k,k'}^{ch,QQ'})_{ll'mm'} = \delta_{lm}\delta_{l'm'} + (\hat{L}_{k,k'}^{ch,QQ'})_{ll'mm'}. \tag{4.66}$$

By replacing $\hat{V}^{ch}$ in Eq. (4.101) with $\hat{V}^{ch,Q}$, we obtain

$$\hat{L}_{k,k'}^{ch,QQ'} \equiv \frac{T}{2} \sum_{ch1,2} a_{ch1,ch2}^{ch} \hat{B}(k-k',p) \left\{ \hat{V}^{ch1,Q}(k-k'+p)\hat{V}^{ch2,Q'}(-p) + {}^{Q\to Q'}_{Q'\to Q} \right\}.$$

The diagrammatic expression of Eq. (4.67) is given in Fig. 4.9a. Note that

$$\hat{L}_{k,k'}^{ch} \approx \sum_{\{Q,Q'\}} \hat{L}_{k,k'}^{ch,QQ'}. \tag{4.67}$$

- We consider AL-type U-VC beyond Migdal-Eliashberg approximation.
- Charge-ch U-VC is enlarged by spin fluctuations $(\chi^s)^2$ near magnetic QCP.
- U-VC can be expressed in multipole basis as well pseudo spin basis.

Here, we show numerical results of U-VC. In this calculation, we use 16×16 $\boldsymbol{k}$-meshes and 128 Matsubara frequencies. In Fig. 4.10a and b, we show the α_S dependence of maximum value of the enhancement factor on the Fermi surface,

$$\hat{\Lambda}^{ch,\max} \equiv \max_{\boldsymbol{k},\boldsymbol{k}'} |\hat{\Lambda}^{ch}_{k,k'}| \quad (\epsilon_n = \epsilon_{n'} = \pi T). \tag{4.68}$$

Note that $\Lambda^{ch,\max}_{ll'mm'} = \Lambda^{ch,\max}_{l'lm'm}$ is satisfied. We plot only charge-ch U-VC ($\hat{\Lambda}^c$) since it is much larger than spin-ch one near the magnetic QCP. In the present numerical study, we put $g = 0$ in $\hat{\Lambda}^{ch}$, since the contribution from χ^c is negligibly smaller than that from χ^s and $\chi^{s\perp}$ [35]. Obtained results show that U-VC work as large enhancement factors for the charge-ch coupling constant ($|\hat{\Lambda}^c| \gg 1$) near the magnetic QCP ($\alpha_S \lesssim 1$). This behavior originates from the relation

$$\hat{\Lambda}^c_{k,k'} \propto \sum_p \hat{\chi}^s(k - k' + p)\hat{\chi}^s(p) + 2\hat{\chi}^{s\perp}(k - k' + p)\hat{\chi}^{s\perp}(p).$$

This is qualitatively similar to d-electron systems without SOI. In conclusion, U-VC in f-electron systems give significant contribution as well as in d-electron systems.

On the other hand, there are some significant differences from d-electron systems. In fact, in the present f-electron system,

- Various orbital components of U-VC are equally enlarged.
- U-VC in f-electron is even larger than that in d-electron systems.

Next, we show the obtained result in Fig. 4.11a–d, and the maximum of multipole-decomposed U-VC is defined as

$$\hat{\Lambda}^{ch,QQ'} \equiv \max_{\boldsymbol{k},\boldsymbol{k}'} |(\hat{\Lambda}^{ch,QQ'}_{k,k'})| \quad (\epsilon_n = \epsilon_{n'} = \pi T). \tag{4.69}$$

We consider only magnetic (=odd rank) multipole ch for Q, Q' since the contributions from electric multipole ch are negligibly small within RPA. Note that $\hat{\Lambda}^{c,QQ'} = 0$ if $\Gamma \neq \Gamma'$ for $Q(Q) \in \Gamma(\Gamma')$ in the present model. Figure 4.11a and b shows the orbital-diagonal component of U-VC given by $\Lambda^{c,QQ'}_{2222}$. It has peak at $(Q, Q') = (T_x, T_x)$. Subsequently, $(Q, Q') = (J_z, T_z), (T_z, T_z), (D_z, D_z)$ are also enlarged. In Fig. 4.11c and d, we show orbital-off-diagonal component given by $\Lambda^{c,QQ'}_{1211}$. It takes

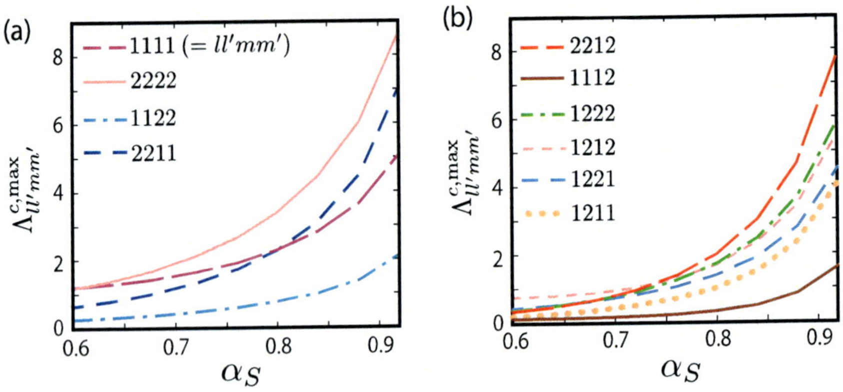

Fig. 4.10 **a**, **b** α_S dependence of charge-channel U-VC. Various orbital components are enlarged near the magnetic QCP ($\alpha_S \lesssim 1$). Reprinted with permission from Ref. [55]. Copyright ©2018 by the American Physics Society

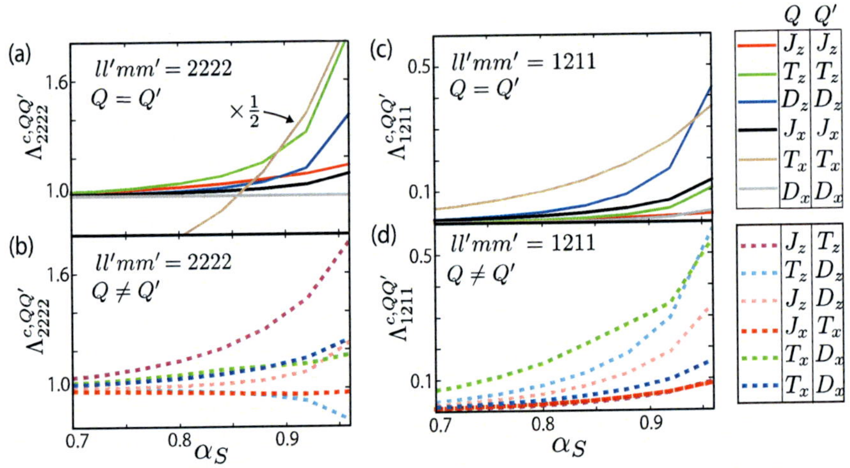

Fig. 4.11 **a**, **c** Obtained U-VC for $Q = Q'$ and **b**, **d** $Q \neq Q'$, higher ranks of multipole fluctuations contribute to the enhancement of U-VC. Reprinted with permission from Ref. [55]. Copyright ©2018 by the American Physics Society

large value for $(Q, Q') = (T_x, D_x)$, (T_z, D_z), (D_z, D_z), (T_x, T_x), (J_z, D_z). Therefore, higher ranks of multipole fluctuations lead to the strong enhancement of U-VC.

In summary, both orbital-diagonal and off-diagonal components of charge-channel U-VC are enlarged. In addition, higher ranks of multipole fluctuations contribute to the enhancement of U-VC. These facts lead to abovementioned differences unlike 3d-electron system. Thus, we conclude that the U-VC in f-electron system plays significant roles due to the strong SOI compared with d-electron systems.

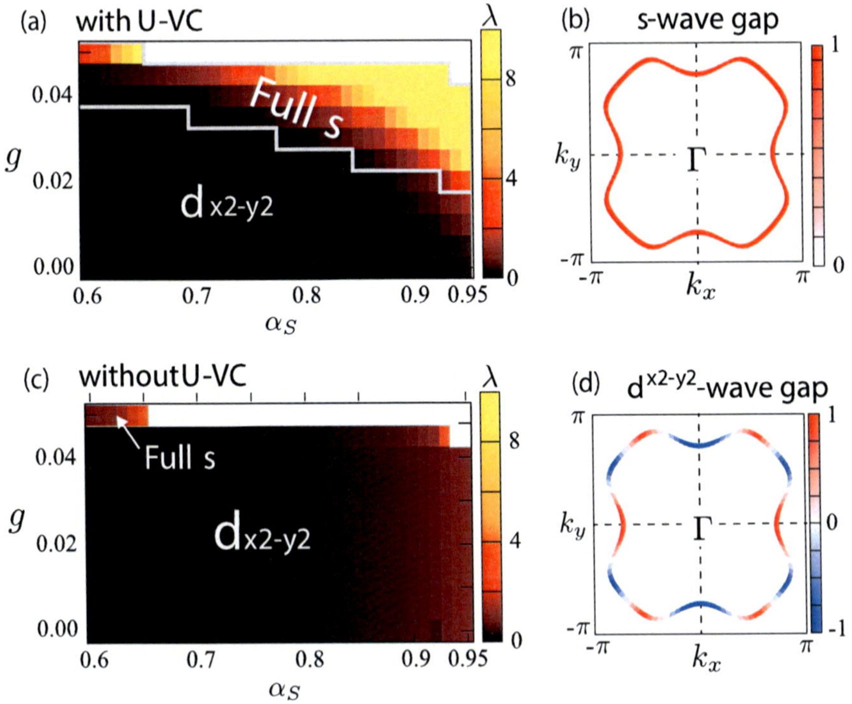

Fig. 4.12 **a** Phase diagram in the presence of U-VC. The s-wave state emerges due to the significant contribution from U-VC. The white region corresponds to $\alpha_C > 1$. **b** The gap function on Fermi surface for s-wave state. **c** Phase diagram in the absence of U-VC. Anisotropic $d_{x^2-y^2}$-wave state appears in wide parameter region. **d** The gap function for $d_{x^2-y^2}$-wave. Reprinted with permission from Ref. [55]. Copyright ©2018 by the American Physics Society

- Charge-channel pairing interaction is enhanced by U-VC.
- Various orbital components of U-VC are equally enlarged.
- Higher ranks of multipole fluctuations enhance U-VC.

SC Phase Diagram

Now, we show obtained superconducting phase diagram by solving the gap equation in Eq. (4.53). We solve the gap equation in the presence of both u and g by the following replacement:

$$u\hat{U} \to u\hat{U} + 2g(0)\mathbf{C}_{A_1^+}\mathbf{C}^\dagger_{A_1^+}. \tag{4.70}$$

Figure 4.12a–b shows the obtained phase diagrams, which is given by the largest eigenvalue and symmetry of eigenvector. In the presence of U-VC, fully gapped s-wave state without any sign reversal emerges when $\alpha_S \lesssim 1$ and $\alpha_C \lesssim 1$ as shown in Fig. 4.12a. The region of s-wave phase gets wider as the magnetic fluctuations develop. These results originate from the fact that the charge-channel attractive interaction is strongly enhanced by the charge-channel U-VC, which is enlarged due to the magnetic (odd-rank) multipole fluctuations when $\alpha_S \lesssim 1$. In fact, charge-channel attractive interaction is expressed as

$$-\frac{1}{2}\hat{V}^c_{U-VC} \propto -\frac{1}{2}|\hat{\Lambda}^c|^2\{\hat{U}^c\hat{\chi}^c\hat{U}^c + \hat{U}^c\}, \tag{4.71}$$

which takes large negative (=attractive) value when $\alpha_C \lesssim 1$ [35]. In addition, we find that quite small $g(0)$ is enough for realizing the s-wave superconductivity. For instance, s-wave state emerges even at $g(0) = 0.025$. This is much smaller than Coulomb interaction $u = 0.31$ at $\alpha_S = 0.9$. Obtained gap functions of s-wave states are almost isotropic as shown in Fig. 4.12b.

In contrast, the s-wave region in Fig. 4.12a is drastically reduced if we neglect U-VC as shown in Fig. 4.12c. In this case, $d_{x^2-y^2}$-wave state appears in wide parameter region. Obtained gap functions of $d_{x^2-y^2}$-wave state are expressed in Fig. 4.12d. It has accidental nodes in addition to the symmetry nodes. Furthermore, the eigenvalue λ for $d_{x^2-y^2}$-wave state in Fig. 4.12c is much smaller than that for s-wave state in Fig. 4.12a, so T_c of $d_{x^2-y^2}$-wave state should be low. We comment that the obtained large eigenvalues λ in Fig. 4.12a, c are overestimated since the self-energy effects (such as the mass renormalization and the quasi-particle damping) are dropped in the gap equation. Then, realistic T_c should be lower than obtained result.

In conclusion, once the small electron-phonon interaction exist, fully gapped s-wave superconducting state appears in f-electron system near the magnetic QCP. This counter-intuitive result is given by the large U-VC caused by multiple (higher rank) multipole fluctuations.

- Fully gapped s-wave SC phase appears near magnetic QCP due to U-VC.
- Small el-ph coupling is enough to realize s-wave SC state ($g(0) \ll u$).

4.3.3 CEF Splitting and f-f Hopping

Here, we discuss CEF splitting ΔE between f_1 and f_2-orbitals, which is given by

$$\Delta E \equiv E_1 - E_2 \tag{4.72}$$

as shown in Fig. 4.13a. Furthermore, we recall that $|f_1\rangle$ and $|f_2\rangle$ have different itinerancy. $|f_1\rangle$ is relatively itinerant while $|f_2\rangle$ is relatively localized. Therefore, obtained DoS at the Fermi level: $D^{f_l}(0)$ behaves as shown in Fig. 4.13b. The ratio

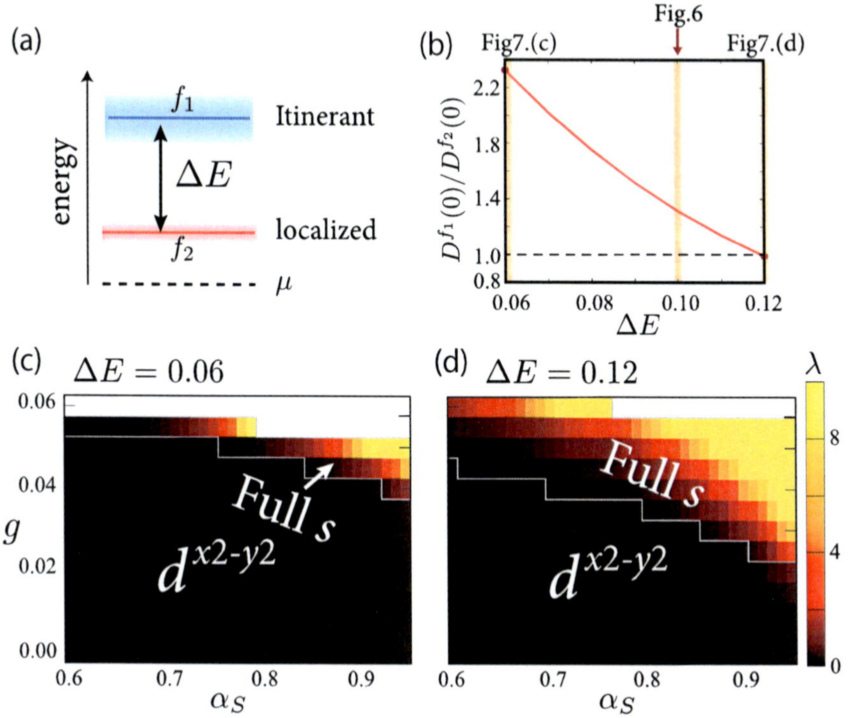

Fig. 4.13 **a** The energy level of the f-orbital states in the present model. **b** ΔE dependence of the ratio of the DoS $D^{f_1}(0)/D^{f_2}(0)$. The ratio goes to unity at $\Delta E \simeq 0.12$. Obtained phase diagram at **c** $\Delta E = 0.06$ and **d** $\Delta E = 0.12$. Reprinted with permission from Ref. [55]. Copyright ©2018 by the American Physics Society

$D^{f_1}(0)/D^{f_2}(0)$ is much larger than unity at $\Delta E = 0$. Then, the ratio decreases with ΔE. At $\Delta E \simeq 0.12$, the ratio reaches unity.

In Fig. 4.13c and d, we show obtained phase diagram at $\Delta E = 0.06$ and $\Delta E = 0.12$, respectively. The region of s-wave state at $\Delta E = 0.12$ is much wider than that at $\Delta E = 0.06$, which means that s-wave state is favored as ΔE increases. As a result, the condition $D^{f_1}(0) \approx D^{f_2}(0)$ is significant for realizing the s-wave superconducting state. In other words, the multi-orbital nature on Fermi surface is important for realizing s-wave states. Therefore, the s-wave state emerges in the presence of finite CEF splitting of f-levels when the s-f hybridization has orbital dependence. This situation is expected to be realized in $CeCu_2Si_2$ at $P = 0$[25].

f-f Hopping

Next, we discuss the effects of f-f hopping. In the previous sections, we neglected f-f hopping. Here, we introduce the orbital-dependent f-f hopping. In this case, f-electron energy E_l have $\boldsymbol{k}$-dependence. As a result, the f_l-orbital weight comes to have θ-dependence on the Fermi surface. The f-f hopping is expressed as

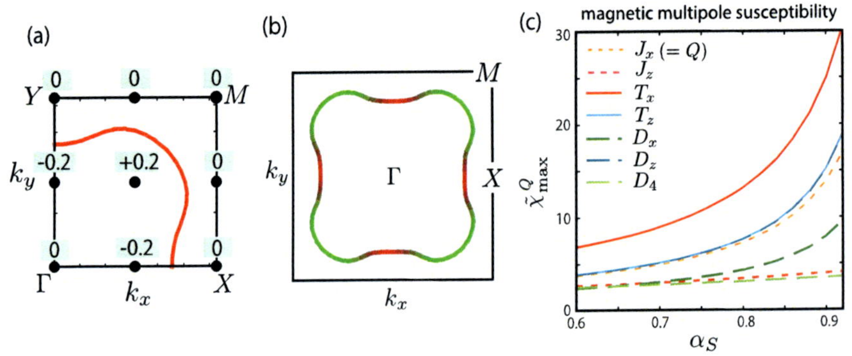

Fig. 4.14 **a** Obtained Fermi surface with f-f hopping. Each number gives intra-orbital energy shift $\delta E_{\boldsymbol{k}}$. **b** f_l-orbital weight on the Fermi surface. The red (green) line corresponds to $f_1(f_2)$-orbital. **c** α_S dependence of magnetic multipole susceptibilities

$$\hat{H}_{ff} = \sum_{kl\sigma} E_{lk} f^{\dagger}_{kl\sigma} f_{kl\sigma}. \quad \left(E_{lk} = E_l - (-1)^l \delta E_k\right). \tag{4.73}$$

Here, $\delta E_{\boldsymbol{k}}$ is given by small f-f hopping integrals: ($|\delta E_{\boldsymbol{k}}| < 0.12|t_{ss}|$) Here, we set $E_{1\boldsymbol{k}} \equiv E_1 + \delta E_{\boldsymbol{k}}$ and $E_{2\boldsymbol{k}} \equiv E_2 - \delta E_{\boldsymbol{k}}$, where the $\boldsymbol{k}$-dependence of $\delta E_{\boldsymbol{k}}$ is shown in Fig. 4.14a. Technically, to realize the $\delta E_{\boldsymbol{k}}$, we introduce the intra-orbital f-f hopping up to fifth nearest neighbor hopping integrals according to Ref. [36]. In Fig. 4.14b, we show the obtained f_l-orbital weight on Fermi surface. It shows strong $\boldsymbol{k}$-dependence irrespective of the fact that $|\delta E_{\boldsymbol{k}}|(\sim 0.2)$ is much smaller than $t_{sf}(= 0.7)$.

One may suspect that higher rank multipole susceptibilities may be suppressed when the f-orbital weight is $\boldsymbol{k}$-dependent, since the orbital off-diagonal components of χ^s may be suppressed. To answer this question, we perform RPA analysis by using PAM model with f-f hopping introduced in Eq. (4.73). Figure 4.14c shows the obtained magnetic multipole susceptibilities. We find that multiple higher rank magnetic multipole susceptibilities develop, which is quite similar to our result without f-f hopping in Fig. 4.5b. This unexpected results originate from the fact that many body effects away from Fermi energy also contribute to the multipole susceptibility. This result strongly indicates that U-VC is still important even in the presence of small f-f hopping.

- s-wave SC phase emerges in the presence of CEF splitting ΔE.
- Multi-orbital nature of Fermi surface is important to realize s-wave SC state (i.e., $D^{f_1}(0) \approx D^{f_2}(0)$).
- Various magnetic multipole fluctuations develop even with f-f hopping.

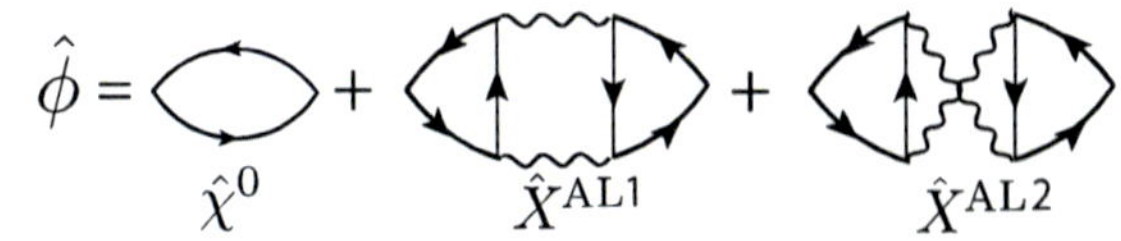

Fig. 4.15 **a** Diagram of irreducible susceptibility with AL-VC

Important Roles of χ-VC

In the previous section, we neglect many body effects in f-electron susceptibilities, which is naturally expected in strongly correlated system. This term is given by three-vertex corrections beyond RPA, which we call χ-VC. Here, we consider the AL-type χ-VC, whose diagrammatic expression is shown in Fig. 4.15. Then, the f-electron susceptibility is given by

$$\hat{\chi}_{\rm AL}(q) = \hat{\phi}(q)(\hat{1} - u\hat{U}\hat{\phi}(q))^{-1},$$
$$\hat{\phi}(q) \equiv \hat{\chi}^0(q) + \hat{X}^{\rm AL}(q).$$

Since χ-VC takes large value only for electric (even-rank) multipole susceptibilities, we project out the magnetic channel contribution of χ-VC. Schematic expression of AL-type χ-VC is given by

$$\hat{X}^{\rm AL}_{QQ'} \propto \sum_{Q_1 \sim Q_4} \hat{V}^{Q_1,Q_2}\hat{V}^{Q_3,Q_4}(\hat{G}^f)^6 \mathbf{O}^Q \mathbf{O}^{Q'\dagger}, \tag{4.74}$$

where $\boldsymbol{k}$-dependence of the matrices is omitted. The detailed expression is written in Appendix 3. Considering the symmetry of the Green function $\hat{G}^f \propto \hat{\sigma}^0$ in Eq. (4.26), finite contribution only comes from

$$\hat{X}^{\rm AL}_{QQ'} \propto {\rm Tr}\{O^Q \cdot O^{Q_1} \cdot O^{Q_4}\} \cdot {\rm Tr}\{O^{Q'} \cdot O^{Q_2} \cdot O^{Q_3}\}. \tag{4.75}$$

In the following numerical study, we set $E_1 = E_2 = 0.1$ and $t^1_{sf} = t^2_{sf} = 0.62$ to make the analysis simple. Then, the relation $D_1(\epsilon) = D_2(\epsilon)$ holds. We adopt $T = 0.045$, $\mu = -0.143$, $n_f = 0.9$, and $n_s = 0.3$. In addition, we neglect el-ph coupling $g(\omega_j) = 0$. Here, we calculate f-electron susceptibilities in IR $\Gamma(= A_1^+, A_2^+, E^+, A_1^-, A_2^-, E^-)$ by solving the following equation:

$$\hat{\chi}_{\rm AL}(\boldsymbol{q}, 0)\mathbf{w}^\Gamma_{\boldsymbol{q}} = \lambda^\Gamma_{\boldsymbol{q}}\mathbf{w}^\Gamma_{\boldsymbol{q}} \qquad \left(\mathbf{w}^\Gamma_{\boldsymbol{q}} = \sum_{Q\in\Gamma} b_Q(\boldsymbol{q})\mathbf{Q}\right), \tag{4.76}$$

where $\mathbf{w}^\Gamma_{\boldsymbol{q}}$ ($\lambda^\Gamma_{\boldsymbol{q}}$) is the eigenvector (eigenvalue) and $b_Q(\boldsymbol{q})$ is real coefficient. Then, the largest eigenvalue gives the multipole susceptibility for Γ.

In Fig. 4.16a, we show obtained susceptibilities for each Γ. With increasing u, all the electric fluctuations strongly develop thanks to the AL-type χ-VC. Thus, large electric susceptibilities originate from the interference of magnetic fluctuations.

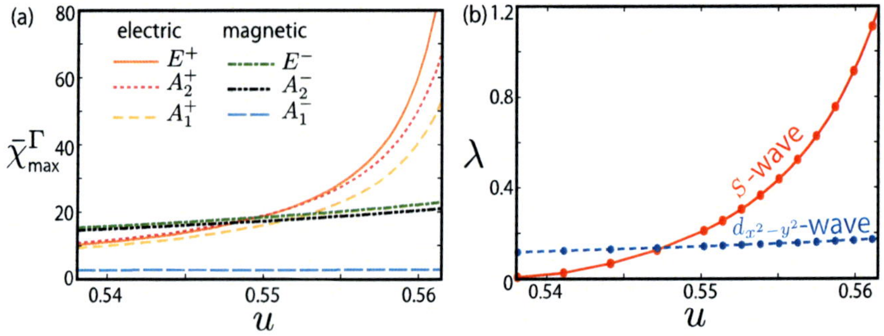

Fig. 4.16 **a** Obtained susceptibility for each IR. Electric susceptibilities ($\Gamma = E^+, A_2^+, A_1^+$) develop due to the AL-VC. **b** Obtained eigenvalue as the function of Coulomb interaction u. The $d_{x^2-y^2}$-wave state is replaced with the fully gapped s-wave state at $u > 0.55$. Reprinted with permission from Ref. [56]. Copyright ©2019 by the Physical Society of Japan

For the electric susceptibilities, the maximum position for $\Gamma = A_1^+$ is $\boldsymbol{q} \approx (\pi, \pi)$, whereas A_2^+, E^+ has peak at $\boldsymbol{q} \approx (0, 0)$. For the magnetic susceptibilities, the maximum position for A_2^-, E^- is $\boldsymbol{q} \approx (\pi/2, \pi/2)$.

Next, we solve the linearized gap equation. In this calculation, we consider both χ-VC and U-VC, while the el-ph coupling is neglected to simplify the discussion. As explained in the previous section, U-VC for the electric channel is enhanced by various ranks of magnetic multipole fluctuations. Therefore, the pairing interaction due to electric fluctuations is strongly enlarged by U-VC. As shown in Fig. 4.16b, $d_{x^2-y^2}$-wave state is replaced with the s-wave one at $u = 0.55$, due to the strong electric fluctuations in Fig. 4.16a. The obtained s-wave state is fully gapped without sign reversal, consistently with experiments in $CeCu_2Si_2$.

- Electric multipole fluctuations in $\Gamma = E^+, A_2^+, A_1^+$ are enhanced by considering χ-VC.
- s-wave SC emerges near AFM phase even in the absence of el-ph coupling.

4.3.4 *Microscopic Origin of s-wave SC*

Now, we discuss the origin of the s-wave superconductivity. First, we introduce the Q, Q'-fluctuation-induced pairing interaction in the band basis $V_{\text{sing}}^{\chi,QQ'}(k, k')$, which is given by the unitary transformation of

$$\hat{V}_{\text{sing}}^{\chi,QQ'}(k,k') \equiv \left\{u^2 \hat{\Lambda}_{k,k'}^{ch} \hat{U}^{ch} \hat{\chi}_{\text{AL}}^{ch,Q,Q'}(k-k') \hat{U}^{ch} \hat{\Lambda}_{-k,-k'}^{ch\dagger}\right\}_{\uparrow\uparrow\downarrow\downarrow} - \{ \qquad \}_{\uparrow\downarrow\uparrow\downarrow} \quad (4.77)$$

Then, SC pairing interaction due to the electric multipole fluctuations is given as

$$V^{\chi,\mathrm{ele}} = \sum_{QQ' \in \mathrm{even}} V^{\chi,QQ'} \left(V^{\chi,QQ'} \equiv \int dkdp V_{\mathrm{sing}}^{\chi,QQ'}(k,p) \right). \quad (4.78)$$

Here, we introduce two different models for further discussion:

$$\delta E_{\boldsymbol{k}} \neq 0 \;\rightarrow \text{Fig.4.14} \quad \text{(model A)} \qquad \delta E_{\boldsymbol{k}} = 0 \quad \text{(model I)}. \quad (4.79)$$

In the model I, the orbital weight is perfectly isotropic, whereas the shape of Fermi surface is almost model independent. In Fig. 4.17a, we show obtained SC pairing interaction in the model A. The contribution from the hexadecapole (H_0) fluctuations in the A_1^+ representation is the largest, while other electric fluctuations are also important. In this case, s-wave SC appears with strong Coulomb interaction. For comparison, Fig. 4.17b shows the result in the orbital isotropic model in the model I. Surprisingly, multipole fluctuations other than H_0 do not contribute to the s-wave pairing, irrespective that all electric multipole susceptibilities (E^+, A_2^+, A_1^+) develop similarly to Fig. 4.16a. In this case, s-wave SC does not appear. Then, obtained fact is summarized as

model A → s-wave SC emerges. model I → Only D-wave SC emerges.

The reason of these nontrivial facts is explained in the next paragraph. The Fermi surface and its orbital character of each model are shown in Fig. 4.17c and d.

In addition, we define the pairing interaction due to both electric and magnetic fluctuations (Coulomb repulsion) as

$$\hat{V}_{\mathrm{sing}}^{\chi}(k,k') \equiv \left\{ u^2 \hat{\Lambda}_{k,k'} \hat{U} \hat{\chi}_{\mathrm{AL}}(k-k') \hat{U} \hat{\Lambda}_{-k,-k'}^{\dagger} \right\}_{\uparrow\uparrow\downarrow\downarrow} - \{ \qquad \}_{\uparrow\downarrow\uparrow\downarrow},$$
$$\hat{V}_{\mathrm{sing}}^{U}(k,k') \equiv \left\{ u^2 \hat{\Lambda}_{k,k'} \hat{U} \hat{\Lambda}_{-k,-k'}^{\dagger} \right\}_{\uparrow\uparrow\downarrow\downarrow} - \{ \qquad \}_{\uparrow\downarrow\uparrow\downarrow}. \quad (4.80)$$

Figure 4.18 shows the obtained interactions with lowest frequency ($\epsilon_n = \epsilon_m = \pi T$) averaged over the Fermi surface. Absolute value of $V_{\mathrm{sing}}^{\chi(U)}$ increases with u due to U-VC for the electric channel. However, in model I, the total pairing interaction $V_{\mathrm{sing}}^{\chi} + V_{\mathrm{sing}}^{U}$ is always negative (=repulsive), so the d-wave state appears. In contrast, in model A, the total pairing interaction becomes positive with u since not only H_0 fluctuations, but also other electric fluctuations contribute to the attractive pairing when $\delta E_{\boldsymbol{k}} \neq 0$. Therefore, fully gapped s-wave state is realized in model A for $u > 0.55$, as shown in Fig. 4.16b. For $u \sim 0.55$, the total pairing interaction is negative at $\epsilon_n = \epsilon_m = \pi T$ as shown in Fig. 4.18e. Nonetheless, the s-wave state is realized because of the retardation effect due to the strong frequency dependence of V^{sing}.

Finally, we discuss why all the electric fluctuations contribute to the s-wave state in model A, while only D-wave state appears in model I. In both model, electric and magnetic fluctuations are similar since $D_1(\epsilon) \approx D_2(\epsilon)$ holds even if $\delta E_{\boldsymbol{k}} \neq 0$. In

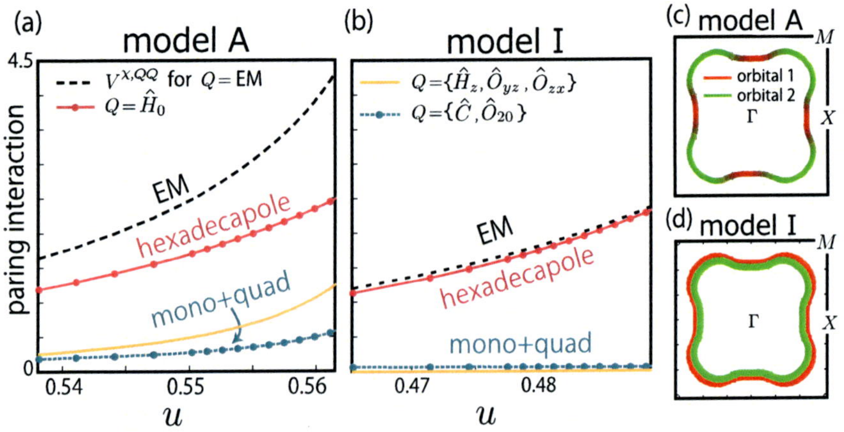

Fig. 4.17 Obtained $V^{\chi,QQ}$ and $V^{\chi,\mathrm{EM}}$ due to the electric multipole in **a** model A and **b** model I. **c**, **d** Orbital character on the Fermi surface. Reprinted with permission from Ref. [56]. Copyright ©2019 by the Physical Society of Japan

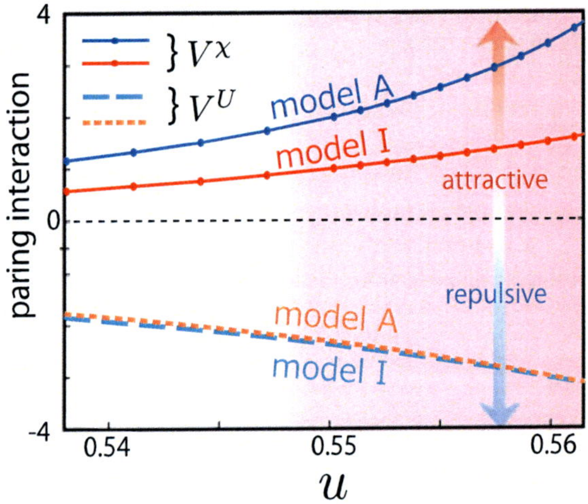

Fig. 4.18 Obtained pairing interactions V^{χ} and V^{U}. For model I, the horizontal axis is shifted by +0.073. Reprinted with permission from Ref. [56]. Copyright ©2019 by the Physical Society of Japan

this situation, one may expect that any electric fluctuation brings attractive pairing interaction. However, some elements of the electric susceptibility $\hat{\chi}^{QQ'}(\boldsymbol{q}, 0)$ can be negative except for monopole, so the cancellation of pairing interaction sometimes occurs. (For example, $\chi_{1212}^{\mathrm{sl}QQ}(\boldsymbol{q}) < 0$ for $Q = O_{yz,zx}$.) Therefore, difference between model A and I is that the "inter-orbital pairing" $\langle f_{\boldsymbol{k}1\uparrow} f_{-\boldsymbol{k}2\downarrow}\rangle$ is suppressed in model A due to the $\boldsymbol{k}$-dependence of the orbital character on the Fermi surface.

Now, we discuss origin of difference between model A and I. The pairing interaction is simplified as

$$\hat{V}^{\mu\nu} = g\mathbf{O}_{\mu\nu}(\mathbf{O}_{\mu\nu})^{\dagger}, \tag{4.81}$$

where $g > 0$ and $\hat{O}_{\mu\nu} \equiv \hat{\sigma}_\mu \hat{\tau}_\nu$ $(\mu, \nu = 0, x, y, z)$. All the even-rank operators are given by linear combination of $\hat{O}_{\mu\nu}$ with $(\mu\nu) = (00), (0x), (0z), (xy), (yy), (zy)$. The gap equation in the orbital basis is given by

$$\lambda\hat{\Delta} \approx T \sum_k g\,'\hat{O}_{\mu\nu}\hat{G}^f(k)\hat{\Delta}\hat{G}^f(-k)\hat{O}_{\mu\nu}\theta(\omega_c - |\epsilon_n|), \tag{4.82}$$

where ω_c is BCS type cutoff. As we mentioned before, $\hat{G}^f$ is expressed as

$$G^f_{lm}(k) = G^{0f}_l(k)\delta_{lm} + (-1)^{l+m}t^2_{sf}(\sin^2 k_y + \sin^2 k_x)G^{0f}_l(k)G^{0f}_m(k)G^c_\uparrow(k) \tag{4.83}$$

We neglect the first term in Eq. (4.83) since it does not give $-\ln T$ term in gap equation. In this case, we obtain

$$\hat{G} \propto \hat{\sigma}_0(\hat{\tau}_0 - a\hat{\tau}_x). \tag{4.84}$$

In addition, $G^{0f}_1(k) = G^{0f}_2(k)$ holds in model I, while $G^{0f}_1(k)G^{0f}_2(k) = 0$ due to $|\delta E_{\boldsymbol{k}}|$ $(\gg t^2_{sf})$ in model A. Therefore, the symmetry of the Green function is approximately given as

$$a = 1 \quad (\text{model A}), a = 0 \quad (\text{model I}). \tag{4.85}$$

Here, the general gap function is written by

$$\hat{\Delta} \propto i\hat{\sigma}_y(\Delta_0\hat{\tau}_0 + \Delta_x\hat{\tau}_x). \tag{4.86}$$

By using Eq. (4.84) and Eq. (4.86), we obtain

$$\hat{G}\hat{\Delta}\hat{G} \propto ((1 + a^2)\Delta_0 - 2a\Delta_x)i\hat{\sigma}_y\hat{\tau}_0 + ((1 + a^2)\Delta_x - 2a\Delta_0)i\hat{\sigma}_y\hat{\tau}_x. \tag{4.87}$$

Therefore, the eigenvalue of the gap equation is given by

$$\lambda = g'(1 + a)^2 \quad \text{for } (\mu\nu) = (00), (0x), \tag{4.88}$$

$$\lambda = g'(1 - a)^2 \quad \text{for } (\mu\nu) = (0z), (xy), (yy), (zy), \tag{4.89}$$

where $g' = gD_1(0)\ln(\omega_c/T)$.

In Fig. 4.19, we summarize the eigenvalue λ for each electric pairing interaction, in the case of $a = 0$ (intra-orbital Cooper pair) and $a = 1$ (intra+inter-orbital Cooper pair). We note that $\hat{O}_{0z} \propto \hat{O}_{20} - 2\hat{C}$ and $\hat{O}_{0x} \propto -3\hat{H}_0 + 2\hat{O}_{20} + \hat{C}$. In case of $a = 0$, all electric fluctuations contribute to the pairing. In case of $a = 1$, however, only $\hat{O}_{0x}$ and C channels contribute to the pairing. In the present PAM, charge (monopole) fluctuations are small, so they do not contribute to the pairing. Since $\hat{O}_{0x}$ is included only in H_0 hexadecapole, the H_0 fluctuations give dominant s-wave pairing interaction. To summarize, the pairing interaction increases if the inter-orbital

model / $\hat{P}_{\mu\nu}$	a=0 (intra orbital CP)	a=1 (inter+intra orbital CP)
	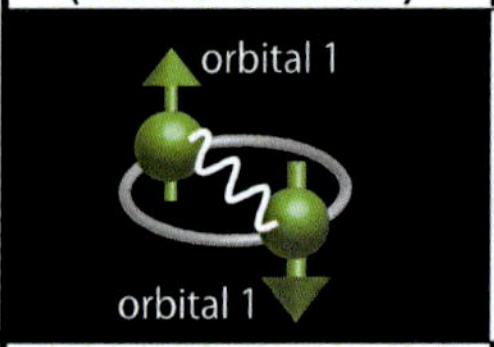	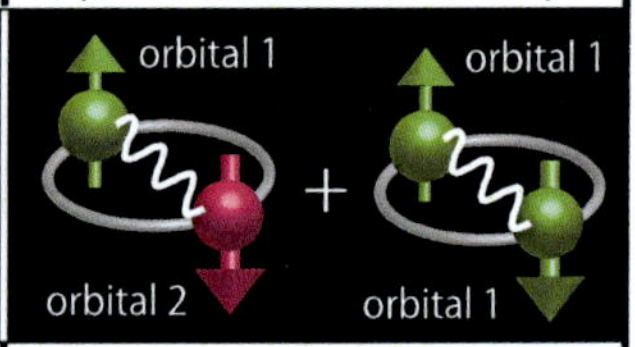
$\hat{P}_{00} \propto \hat{C}$	g	4g
$\hat{P}_{0z} \propto \hat{O}_{20} - 2\hat{C}$	g	0
$\hat{P}_{0x} \propto 3\hat{H}_0 + 2\hat{O}_{20} + \hat{C}$	g	4g
$\hat{P}_{\mu y} \propto \hat{O}_{zx}, \hat{O}_{yz}, \hat{H}_z$	g	0

Fig. 4.19 Eigenvalue λ due to $\hat{O}_{\mu\nu}$ electric interaction for $a = 0$ (intra-orbital Cooper pair) and $a = 1$ (intra+inter-orbital pair). Except for $\hat{O}_{0x}$ and $\hat{C} = \hat{O}_{00}$, the electric fluctuations give repulsive interaction for inter-orbital Cooper pair. Reprinted with permission from Ref. [56]. Copyright ©2019 by the Physical Society of Japan

Cooper pairs are killed by finite $|\delta E_{\boldsymbol{k}}|$, so the numerical results in Fig. 4.18 are well understood.

We also analyzed more realistic PAM with $E_1^f > E_2^f$ and $t_{sf}^1 > t_{sf}^1$. Even in this case, we obtain H_0 hexadecapole fluctuation-mediated fully gapped s-wave state. Thus, our result discussed in this section will be robust.

- s-wave SC state appears in the model A, while only d-wave SC state appears in the model I.
- Dominant attraction comes from the hexadecapole $\hat{H}_0$ fluctuation.
- All electric fluctuations contribute to the s-wave SC pairing.

4.3.5 Retardation Effect

Here, we discuss important roles of retardation effects. In Fig. 4.20a, we show the obtained pairing interaction on the Fermi surface defined as

$$V_{\max}^{\text{sing}}(\omega_j) \equiv \max_{\boldsymbol{k},\boldsymbol{k}'} \left\{ V^{\text{sing}}(\boldsymbol{k}, \pi T, \boldsymbol{k}', \pi T + \omega_j) \right\}. \tag{4.90}$$

The pairing interaction is attractive (positive) at $\omega_j = 0$, whereas it becomes to repulsion for $\omega_j > 0$. For this reason, the gap function defined as $\Delta(\epsilon_n) \equiv \max_{\boldsymbol{k}}\{\Delta(\boldsymbol{k}, \epsilon_n)\}$

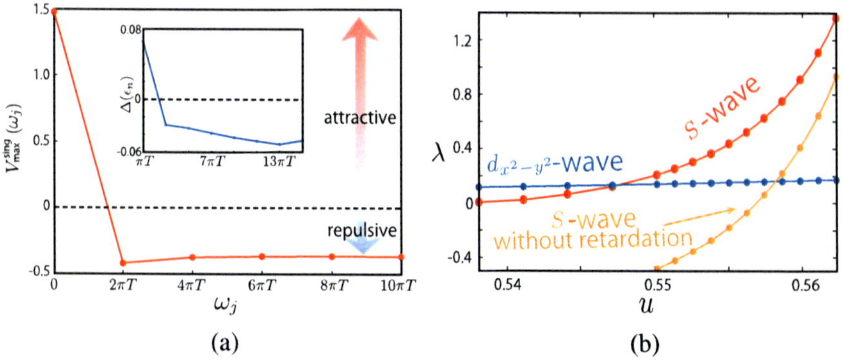

Fig. 4.20 **a** Obtained pairing interaction $V_{\max}^{\text{sing}}(\omega_j)$ and gap function $\Delta(\epsilon_j)$ (inset) as the function of Matsubara frequency. Strong retardation effect is recognized. **b** Obtained eigenvalue as the function of u. The eigenvalue of the s-wave state is significantly enlarged by the retardation effect

shows the sign change as the function of ϵ_n, as shown in the inset of Fig. 4.20. This is a hallmark of the retardation effects due to the strong ω_j-dependence of the electric (even-rank) fluctuation. The smiler result should be obtained when we consider the cutoff energy given by el-ph interaction.

Due to the retardation effect, the direct Coulomb depairing potential is reduced from U to

$$U^* \sim \frac{U}{1 + UD(0)\ln(E_F/\omega_0)}, \tag{4.91}$$

where ω_0 is the energy scale of the electric fluctuations. Since $\omega_0 \ll E_F$, the fully gapped s-wave superconductivity can be stabilized in HF systems.

To clarify the importance of the retardation effect, we calculate the gap equation for the following pairing interaction only at $|\epsilon_n| = |\epsilon_m| = \pi T$:

$$\tilde{V}^{\text{sing}}(\boldsymbol{k}, \boldsymbol{k}') \equiv V^{\text{sing}}(\boldsymbol{k}, \pi T, \boldsymbol{k}', \pi T). \tag{4.92}$$

Then, the reduction of the depairing potential ($U \to U^*$) is not taken into account.

Figure 4.20b shows the eigenvalue for the s-wave state "without retardation effect", derived from Eq. (4.92). For comparison, we also plot the eigenvalues derived from the frequency-dependent $V^{\text{sing}}(k, k')$. The eigenvalue for the s-wave state derived from $\tilde{V}^{\text{sing}}(\boldsymbol{k}, \boldsymbol{k}')$ is smaller than that derived from $V^{\text{sing}}(k, k')$. Thus, the eigenvalue for the s-wave state is significantly enlarged by the reiteration effect ($U \to U^*$).

- Retardation effect is important to explain s-wave SC phase.
- Energy dependence of electric (el-ph) fluctuations gives retardation effect.

4.4 Summary

In this section, we proposed a microscopic origin of fully gapped s-wave superconductivity in multi-orbital HF systems beyond ME formalism. In the present $\Gamma_7^{(1)}$-$\Gamma_7^{(2)}$ quartet PAM, various magnetic multipole fluctuations develop, due to the cooperation between strong SOI and Coulomb interaction as shown in Fig. 4.5. These fluctuations give significant U-VC in SC pairing interaction. Also, we verified that our present result does not change qualitatively compared with those for $\Delta E = 0 \sim 0.2$. In particular, attractive pairing interaction of electric ch is prominently enhanced by U-VC as plotted in Figs. 4.10 and 4.11. As a result, even-rank multipole fluctuations bring attractive interaction when the system approaches to AFM-QCP. T_c of s-wave SC state is strongly enhanced near AFM-QCP in multi-orbital HF systems. In addition, it comes to be easier to obtain s-wave SC phase when the moderate phonon-induced multipole fluctuations exist as shown in Fig. 4.12. The present mechanism may be responsible for the fully gapped s-wave superconducting state realized in $CeCu_2Si_2$.

Here, we comment that el-ph coupling is expected in HF systems. In fact, large Gruneisen parameter ($\eta \equiv -d\log T_K / d\log\Omega \sim 30 - 80$) was observed, which means el-ph interaction is significant [50]. Idea of phonon-mediated s-wave superconductivity in HF systems has been discussed in Refs. [50–52]. Now, this scenario becomes more realistic by considering U-VC.

Another possible origin of electric fluctuations is many body effects due to AL-type χ-VC. By considering χ-VC, electric fluctuations drastically develop. After that, attractive SC pairing interaction is enhanced by large U-VC. Based on the present model, H_0 hexadecapole fluctuations dominate over other even-rank fluctuations as a attraction for SC Cooper pairs. Thus, s-wave state is caused by the hexadecapole (rank 4) fluctuations. Also, s-wave SC phase is stabilized by other electric fluctuations including quadrupole and monopole, when we are introducing small δE_{kl}, In this case, the inter-orbital Cooper pairs are killed. Therefore, we revealed that all of χ-VC, U-VC, and e-ph interactions enhance T_c of s-wave SC, cooperatively. The present pairing mechanism is expected to be important to understand various HF materials.

The main results of the present study are summarized as follows:

- Near the magnetic QCP, several higher rank multipole fluctuations strongly develop simultaneously, whereas dipole one solely develops in d-electron systems.
- Moderate electric multipole fluctuations develop due to cooperation of the el-ph coupling and many body effects beyond RPA (χ-VC).
- Development of magnetic multiple multipole fluctuations gives prominent U-VC, which leads to the violation of ME scheme.
- Owning to U-VC, electric-multipole-fluctuation-induced s-wave SC phase is stabilized when $D^{f1}(0) \approx D^{f2}(0)$. This is a necessary condition for realizing moderate quadrupole or hexadecapole susceptibility.

Also, there remain unsolved future issues. For instance, renormalization of self-energy, which brings strong mass enhancement, is one of the important issues. Also, pressure-induced second SC dome in $CeCu_2Si_2$ is uncovered problem [27]. In addition, we expect that the present mechanism may be applicable for triplet SC phase in UPt_3.

Appendix 3

Here, we explain about the ground states of $J = 5/2$ PAM and derivation of complex hopping parameters given by Eq. (4.4) based on Slater-Koster table. The ground states in the present model in l_z-basis are written as

$$|f_1 \Downarrow\rangle = a\left\{\sqrt{\frac{1}{7}}|2, \uparrow\rangle - \sqrt{\frac{6}{7}}|3, \downarrow\rangle\right\} + b\left\{\sqrt{\frac{5}{7}}|-2, \uparrow\rangle - \sqrt{\frac{2}{7}}|-1, \downarrow\rangle\right\}, \quad (4.93)$$

$$|f_1 \Uparrow\rangle = a\left\{\sqrt{\frac{6}{7}}|-3, \uparrow\rangle - \sqrt{\frac{1}{7}}|-2, \downarrow\rangle\right\} + b\left\{\sqrt{\frac{2}{7}}|1, \uparrow\rangle - \sqrt{\frac{5}{7}}|2, \downarrow\rangle\right\}, \quad (4.94)$$

$$|f_2 \Uparrow\rangle = -a\left\{\sqrt{\frac{2}{7}}|1, \uparrow\rangle - \sqrt{\frac{5}{7}}|2, \downarrow\rangle\right\} + b\left\{\sqrt{\frac{6}{7}}|-3, \uparrow\rangle - \sqrt{\frac{1}{7}}|-2, \downarrow\rangle\right\}, \quad (4.95)$$

$$|f_2 \Downarrow\rangle = -a\left\{\sqrt{\frac{5}{7}}|-2, \uparrow\rangle - \sqrt{\frac{2}{7}}|-1, \downarrow\rangle\right\} + b\left\{\sqrt{\frac{1}{7}}|2, \uparrow\rangle - \sqrt{\frac{6}{7}}|3, \downarrow\rangle\right\}, \quad (4.96)$$

where $\Uparrow$ ($\Downarrow$) is pseudo-spin and $\uparrow$ ($\downarrow$) denotes the real spin. Note that the wave functions for $l_z = \pm 2$ are proportional to z,

$$\langle \mathbf{r} \mid \pm 2, \sigma\rangle \propto z.$$

Therefore, in 2D system, the hopping integrals between s-orbital and $l_z = 2$ electrons should be zero. Thus, $\langle s, \sigma, \mathbf{R}_i | \pm 2, \sigma, \mathbf{R}_j\rangle = 0$ is satisfied. Finally, we obtain

$$\langle s \uparrow |f_1 \Uparrow\rangle = \sqrt{\frac{6}{7}}\langle s \uparrow | -3, \uparrow\rangle, \quad \langle s \downarrow |f_1 \Downarrow\rangle = -\sqrt{\frac{6}{7}}\langle s \downarrow |3, \downarrow\rangle, \quad (4.97)$$

$$\langle s \uparrow |f_2 \Uparrow\rangle = -\sqrt{\frac{2}{7}}\langle s \uparrow |1, \uparrow\rangle, \quad \langle s \downarrow |f_2 \Downarrow\rangle = \sqrt{\frac{2}{7}}\langle s \downarrow | -1, \downarrow\rangle, \quad (4.98)$$

where $a = 1, b = 0$ is adopted. Here, each wave function $|\pm 3, \sigma\rangle, |\pm 1, \sigma\rangle$ is written in (x, y, z)-basis.

$$|\pm 3, \sigma\rangle = -\sqrt{\frac{5}{16}}\left\{\pm x(5x^2 - 3r^2) - iy(5y^2 - 3r^2)\right\} - \sqrt{\frac{3}{16}}\left\{\mp x(y^2 - z^2) - iy(z^2 - x^2)\right\},$$

$$|\pm 1, \sigma\rangle = -\sqrt{\frac{3}{16}}\left\{\mp x(5x^2 - 3r^2) - iy(5y^2 - 3r^2)\right\} - \sqrt{\frac{5}{16}}\left\{\pm x(y^2 - z^2) - iy(z^2 - x^2)\right\}.$$

Therefore, we obtain the hopping parameters:

$$
\begin{aligned}
\langle s| \pm 3, \sigma\rangle &= -\sqrt{\frac{5}{16}}\left\{\pm\langle s|x(5x^2-3r^2)\rangle\langle s|y(5y^2-3r^2)\rangle\right\} \\
&\quad -\sqrt{\frac{3}{16}}\left\{\mp\langle s|x(y^2-z^2)\rangle - i\langle s|y(z^2-x^2)\rangle\right\}, \\
&= -\sqrt{\frac{5}{16}}\left\{\pm\frac{l}{2}(5l^2-3) - i\frac{m}{2}(5m^2-3)\right\} \\
&\quad -\sqrt{\frac{3}{16}}\left\{\mp\frac{\sqrt{15}}{2}l(m^2-n^2) - i\frac{\sqrt{15}}{2}m(n^2-l^2)\right\} t_{sf}.
\end{aligned}
$$

$$
\begin{aligned}
\langle s| \pm 1, \sigma\rangle &= -\sqrt{\frac{3}{16}}\left\{\mp\langle s|x(5x^2-3r^2)\rangle - i\langle s|y(5y^2-3r^2)\rangle\right\} \\
&\quad -\sqrt{\frac{5}{16}}\left\{\pm\langle s|x(y^2-z^2)\rangle - i\langle s|y(z^2-x^2)\rangle\right\}, \\
&= -\sqrt{\frac{3}{16}}\left\{\mp\frac{l}{2}(5l^2-3) - i\frac{m}{2}(5m^2-3)\right\} \\
&\quad -\sqrt{\frac{5}{16}}\left\{\pm\frac{\sqrt{15}}{2}l(m^2-n^2) - i\frac{\sqrt{15}}{2}m(n^2-l^2)\right\} t_{sf},
\end{aligned}
$$

where $l = \cos\phi\sin\theta, m = \sin\phi\sin\theta, andn = \cos\theta$. Here, $t_{sf} = (sf\sigma)$. Then, we consider the nearest neighbor hopping on 2D square lattice, that is, Ce-ions at $(l, m, n) = (0, 1, 0), (-1, 0, 0), (0, -1, 0), (1, 0, 0)$. Therefore, we obtain

$$
\begin{aligned}
\langle s| \pm 3, \sigma\rangle_{NN} &= \sqrt{\frac{5}{16}}\left\{ie^{ik_ya} \pm e^{-ik_xa} - ie^{-ik_ya} \mp e^{ik_xa}\right\} t_{sf} \\
&= \sqrt{\frac{5}{16}}(-2\sin k_y \mp 2i\sin k_x)t_{sf},
\end{aligned} \tag{4.99}
$$

$$
\begin{aligned}
\langle s| \pm 1, \sigma\rangle_{NN} &= \sqrt{\frac{3}{16}}\left\{ie^{ik_ya} \mp e^{-ik_xa} - ie^{-ik_ya} \pm e^{ik_xa}\right\} t_{sf} \\
&= \sqrt{\frac{3}{16}}(-2\sin k_y \pm 2i\sin k_x)t_{sf}.
\end{aligned} \tag{4.100}
$$

Then, the s-f hopping term is given by

$$\langle s\uparrow|f_1\Uparrow\rangle=-\sqrt{\frac{15}{14}}t_{sf}(\sin k_y-i\sin k_x),\quad \langle s\downarrow|f_1\Downarrow\rangle=\sqrt{\frac{15}{14}}t_{sf}(\sin k_y+i\sin k_x),$$
$$\langle s\uparrow|f_2\Uparrow\rangle=\sqrt{\frac{3}{14}}t_{sf}(\sin k_y-i\sin k_x),\quad \langle s\downarrow|f_2\Downarrow\rangle=-\sqrt{\frac{3}{14}}t_{sf}(\sin k_y+i\sin k_x).$$

Also, we introduce the analytic expressions for dressed four-point vertex in χ-VC without using spin and charge channel to study f-electron system with strong SOI. First, the expression for the AL1-term is given as

$$U^{\mathrm{AL1}}_{\alpha\beta}(q)=\frac{T}{2}\sum_{\alpha'\alpha''\beta'\beta''p}C^{\alpha}_{\alpha'\beta''}(q,p)I_{\alpha'\beta'}(p-q)I_{\alpha''\beta''}(p)C^{\beta*}_{\beta'\alpha''}(\bar{q},\bar{p}),\quad(4.101)$$

where $p\equiv(\boldsymbol{p},\omega_j)$, $\bar{p}\equiv(\boldsymbol{p},-\omega_j)$. The three-point function in Eq. (4.101) is given as

$$C^{EF}_{ABCD}(q,p)\equiv-T\sum_{k}G^{f}_{AF}(k-q)G^{f}_{EC}(k)G^{f}_{DB}(k-p),\qquad(4.102)$$

where $\hat{G}^f$ is the f-electron Green function. Also, the expression for the AL2-term in χ-VC is given as

$$U^{\mathrm{AL2}}_{\alpha\beta}(q)=\frac{T}{2}\sum_{\alpha'\beta'\alpha''\beta''p}C'^{\alpha}_{\alpha'\beta''}(q,p)I_{\beta''\beta'}(p-q)I_{\alpha''\alpha'}(p)\tilde{C}'^{\beta}_{\alpha''\beta'}(q,p),\ (4.103)$$

where

$$C'^{EF}_{ABCD}(q,p)\equiv-T\sum_{k}G^{f}_{BF}(k-q)G^{f}_{ED}(k)G^{f}_{CA}(k-q+p),$$
$$\tilde{C}'^{EF}_{ABCD}(q,p)\equiv-T\sum_{k}G^{f}_{AE}(k+q)G^{f}_{FC}(k)G^{f}_{DB}(k+q-p).$$

The total χ-VC is given by $\hat{U}^{\mathrm{AL}}=\hat{U}^{AL1}+\hat{U}^{AL2}$ by eliminating the double counting due to the 2n order term $\propto u^2$. Then, U-VC in f-electron system is also given by

$$(\hat{U}_{kk'})_{LL'MM'}=\frac{T}{2}\sum_{p,A\sim F}B^{MM'}_{ABCDEF}(k-k',p,k')I_{LACD}(k-k'+p)I_{BL'EF}(-p),$$

where

$$B^{MM'}_{ABCDEF}(q,p,k')=G^{f}_{AB}(k'-p)\left\{C''^{MM'}_{CDEF}(q,p)+C''^{MM'}_{EFCD}(q,q+p)\right\}(4.104)$$

$$C''^{AB}_{CDEF}(q,p) \equiv -T \sum_{k'} G^f_{CA}(k'+q) G^f_{BF}(k') G^f_{ED}(k'-p). \tag{4.105}$$

In addition, the expression for the AL2-term in U-VC is given as

$$U^{\mathrm{AL2}}_{\alpha\beta}(q) = \frac{T}{2} \sum_{\alpha'\beta'\alpha''\beta''} \Lambda^{\alpha}_{\alpha'\beta''}(q,p) I_{\beta''\beta'}(p-q) I_{\alpha''\alpha'}(p) \tilde{\Lambda}^{\beta}_{\alpha''\beta'}(q,p), \tag{4.106}$$

where

$$\Lambda^{EF}_{ABCD}(q,p) \equiv -T \sum_{k} G^f_{BF}(k-q) G^f_{ED}(k) G^f_{CA}(k-q+p),$$
$$\tilde{\Lambda}^{EF}_{ABCD}(q,p) \equiv -T \sum_{k} G^f_{AE}(k+q) G^f_{FC}(k) G^f_{DB}(k+q-p).$$

The expression for the MT-term is written as

$$U^{MT}_{LL'MM'}(q) = T^2 \sum_{p,k,A\sim D} G_{LA}(k+q-p) G_{BL'}(k-p) G_{DM}(k+q) G_{M'C}(k) I_{DACB}(p).$$

The total U-VC is given by $\hat{U}^{\mathrm{AL+MT}} = \hat{U}^{AL1} + \hat{U}^{AL2} + \hat{X}^{MT}$, by subtracting the double counting in the same way of χ-VC.

References

1. T. Moriya, K. Ueda, Adv. Phys. Fig. **49**, 555 (2000)
2. Y. Yanase, T. Jujo, T. Nomura, H. Ikeda, T. Hotta, K. Yamada, Phys. Rep. **387**, 1 (2003)
3. H. Kontani, Rep. Prog. Phys. **71** (2008)
4. P. Monthoux, D. Pines, G.G. Lonzarich, Nature **450**, 1177 (2007)
5. T. Dahm, L. Tewordt, Phys. Rev. B **52**, 1297 (1995)
6. P. Monthoux, D.J. Scalapino, Phys. Rev. Lett. **72**, 1874 (1994)
7. T. Takimoto, T. Hotta, K. Ueda, J. Phys.: Condens. Matter **15**, S2087 (2003)
8. R. Shiina, S. Shiba, T. Peter, J. Phys. Soc. Jpn. **66**, 1741 (1997)
9. H. Kusunose, Y. Kuramoto, J. Phys. Soc. Jpn. **74**, 3139 (2005)
10. T. Takimoto, J. Phys. Soc. Jpn. **75** (2006)
11. K. Haule, G. Kotliar, Nat. Phys. **5**, 796 (2009)
12. R. Okazaki, T. Shibauchi, H.J. Shi, Y. Haga, T.D. Matsuda, E. Yamamoto, Y. Onuki, H. Ikeda, Y. Matsuda, Science **331**, 439 (2011)
13. H. Ikeda, M.-T. Suzuki, R. Arita, T. Takimoto, T. Shibauchi, Y. Matsuda, Nat. Phys. **8**, 528 (2012)
14. K. Izawa, H. Yamaguchi, Y. Matsuda, H. Shishido, R. Settai, Y. Onuki, Phys. Rev. Lett. **87** (2001)
15. T. Onimaru, K.T. Matsumoto, Y.F. Inoue, K. Umeo, Y. Saiga, Y. Matsushita, R. Tamura, K. Nishimoto, I. Ishii, T. Suzuki1, T. Takabatake, J. Phys. Soc. Jpn. **7**(9), 033704 (2010)
16. A. Sakai, S. Nakatsuji, J. Phys. Soc. Jpn. **80** (2011)

17. F. Steglich, J. Aarts, C.D. Bredl, W. Lieke, D. Meschede, W. Franz, H. Schafer, Phys. Rev. Lett. **43**, 1892 (1979)
18. H.Q. Yuan, F.M. Grosche, M. Deppe, C. Geibel, G. Sparn, F. Steglich, Science **302**, 21042107 (2003)
19. C. Peiderer, Rev. Mod. Phys. **81**, 1551 (2009)
20. K. Ishida, Y. Kawasaki, K. Tabuchi, K. Kashima, Y. Kitaoka, K. Asayama, C. Geibel, F. Steglich, Phys. Rev. Lett. **82**, 5353 (1999)
21. S. Kittaka, Y. Aoki, Y. Shimura, T. Sakakibara, S. Seiro, C. Geibel, F. Steglich, H. Ikeda, K. Machida, Phys. Rev. Lett. **112** (2014)
22. S. Kittaka, Y. Aoki, Y. Shimura, T. Sakakibara, S. Seiro, C. Geibel, F. Steglich, Y. Tsutsumi, H. Ikeda, K. Machida, Phys. Rev. B **94** (2016)
23. T. Yamashita, T. Takenaka, Y. Tokiwa, J.A. Wilcox, Y. Mizukami, D. Terazawa, Y. Kasahara, S. Kittaka, T. Sakakibara, M. Konczykowski, S. Seiro, H.S. Jeevan, C. Geibel, C. Putzke, T. Onishi, H. Ikeda, A. Carrington, T. Shibauchi, Y. Matsuda, Sci. Adv. **3** (2017)
24. G. Pang, M. Smidman, J. Zhang, L. Jiao, Z. Weng, E.M. Nica, Y. Chen, W. Jiang, Y. Zhang, W. Xie, H.S. Jeevan, H. Lee, P. Gegenwart, F. Steglich, Q. Si, H. Yuan, Proc. Natl. Acad. Sci. USA **115**, 5343 (2018)
25. L.V. Pourovskii, P. Hansmann, M. Ferrero, A. Georges, Phys. Rev. Lett. **112** (2014)
26. K. Hattori, J. Phys. Soc. Jpn. **79** (2010)
27. A.T. Holmes, D. Jaccard, K. Miyake, Phys. Rev. B **69** (2004)
28. H. Ikeda, M.-T. Suzuki, R. Arita, Phys. Rev. Lett. **114** (2015)
29. S. Onari, H. Kontani, Phys. Rev. Lett. **109** (2012)
30. S. Onari, Y. Yamakawa, H. Kontani, Phys. Rev. Lett. **116** (2016)
31. Y. Yamakawa, S. Onari, H. Kontani, Phys. Rev. X **6** (2016)
32. J.R. Schrieffer, J. Low Temp. Phys. **99**, 314 (1995)
33. M.H. Sharifzadeh Amin, P.C.E. Stamp, Phys. Rev. Lett **7**(7), 3017 (1996)
34. A.B. Migdal, J. Exp. Thore. Phys. **34**, 1438 (1958)
35. R. Tazai, Y. Yamakawa, M. Tsuchiizu, H. Kontani, J. Phys. Soc. Jpn. **86** (2017)
36. Y. Yamakawa, H. Kontani, Phys. Rev. B **96** (2017)
37. A. Georges, G. Kotliar, W. Krauth, M.J. Rozenberg, Rev. Mod. Phys. **68**, 13 (1996)
38. G. Kotliar, D. Vollhardt, Phys. Today **57**, 53 (2004)
39. K. Held, Adv. Phys. **56**, 829 (2007)
40. J.H. Shim, K. Haule, G. Kotliar, Sci. Sci. **318**, 1615 (2007)
41. J. Otsuki, Phys. Rev. Lett. **115** (2015)
42. K. Takegahara, Y. Aoki, A. Yanase, J. Phys. C **13**, 583 (1980)
43. T. Saso, H. Harima, J. Phys. Soc. Jpn. **72**, 1131 (2003)
44. A. Amorese, N. Caroca-Canales, S. Seiro, C. Krellner, G. Ghiringhelli, N.B. Brookes, D.V. Vyalikh, C. Geibel, K. Kummer, arxiv:1803.11068
45. J.C. Slater, Phys. Rev. **34**, 1293 (1929)
46. M.R. Norman, Phys. Rev. B **52**, 1421 (1995)
47. T. Inui, Y. Tanabe, Y. Onodera, *Group Theory and Its Applications in Physics*, Springer, Berlin (1990), ISBN 978-3-642-80021-4. https://www.springer.com/gp/book/9783540604457
48. H. Kontani, S. Onari, Phys. Rev. Lett. **104** (2010)
49. T. Saito, Y. Yamakawa, S. Onari, H. Kontani, Phys. Rev. B **92** (2015)
50. H. Razafimandimby, P. Fulde, J. Keller, Z. Phys. B **54**, 111 (1984)
51. F.J. Ohkawa, J. Phys. Soc. Jpn. **56**, 713 (1987)
52. K. Miyake, T. Matsuura, H. Jichu, Y. Nagaoka, Prog. Theor. Phys. **72**, 713 (1984)
53. R. Tazai, Y. Yamakawa, M. Tsuchiizu, H. Kontani, Phys. Rev. B **94** (2016)
54. M. Tsuchiizu, Y. Yamakawa, S. Onari, Y. Ohno, H. Kontani, Phys. Rev. B **91** (2015)
55. R. Tazai, H. Kontani, Phys. Rev. B **98** (2018)
56. R. Tazai, H. Kontani, J. Phys. Soc. Jpn. **88**(6) (2019)

Chapter 5
Multipole Phase

5.1 Background

Heavy fermion systems cause interesting phenomena originating from exotic electronic states with strong Coulomb interaction and SOI in f-electrons. In particular, it is well known that AFM fluctuations cause interesting quantum transition and SC phase [1–8]. In the case of HF system, various multipole order and fluctuations appear, which is absent in transition metal oxides. One of the typical multipole-ordered systems is CeB_6. It shows AF-quadrupole order with $\boldsymbol{q} = (\pi, \pi, \pi)$ around $T_Q = 3.2$K. Then, magnetic dipole order appears at $T_N = 2.4$K [9–12]. Moreover, AF-octupole order is induced in small magnetic field [13–16]. Therefore, it is expected that various multipole fluctuations appear simultaneously in the $H - T$ phase diagram of CeB_6.

Until now, CeB_6 has been studied intensively based on localized f-electron picture [13–19]. However, recent ARPES and neutron inelastic scattering for CeB_6, as well as dHvA for $Ce_xLa_{1-x}B_6$, uncovered the itinerant nature of the f-electron system above $T \sim T_Q$[20]-[25]. These findings mean that itinerant property becomes a reasonable starting point to study the multipole physics of CeB_6. In this section, we study the long standing problem by considering χ-VC.

If we perform RPA study on the PAM, quadrupole order will not be obtained. In particular, odd-rank (=magnetic) multipole fluctuations emerge, while even-rank (=electric) ones remain small within RPA [24, 26, 27]. This fact originates from the importance of VC, due to the many body effects omitted in RPA. For a long time, Fermi liquid approach has been succeeded in HF materials, such as CeB_6 [24], URu_2Si_2 [26], and $CeCu_2Si_2$ [27]. In this case, large Coulomb interaction is renormalized into zU. In many HF systems, $z \ll 1$ is satisfied. Therefore, Fermi liquid theory is applicable even if Coulomb repulsion is quite large.

In the previous study, the lowest ordered VC with respect to fluctuations, called MT-terms, has been studied. Due to the MT-term, rank-5 multipole-ordered state is stabilized in URu_2Si_2 [26]. On the other hand, the MT-VC did not affect even-rank multipole fluctuations. Thus, microscopic origin of quadrupole order observed in various compounds remains as an important issue. For this purpose, CeB_6 is suitable to reveal a theoretical origin of multipole order in HF systems.

R. Tazai, *Theory of Multipole Fluctuation Mediated Superconductivity and Multipole Phase*, Springer Theses,
https://doi.org/10.1007/978-981-16-1026-4_5

Recently, it has been revealed that AL-VC becomes more important than MT-VC, AL-VC works a trigger for realizing nematic order in Fe-based superconductors [28–30]. Analytically, AL-VC is related to the magnetic correlation length ξ as ξ^{4-d} in d-dimension systems. Thus, AL-VC plays significant roles near the AFM-QCP, which is verified by fRG study with higher order VC in an unbiased way [31]-[37]. In the previous section, we revealed the important role of VC in f-electron system [27, 38]. This fact indicates significance of AL-VC in multipole HF systems, which has not been studied so far.

5.2 Formalism

Γ_8 Periodic Anderson Model

Here, we introduce 2D PAM for describing CeB_6. For f-electron states, we consider the Γ_8 quartet in $J = 5/2$ space due to the strong SOI [13]:

$$
\begin{aligned}
|f_1 \Downarrow (\Uparrow)\rangle &= \sqrt{\frac{5}{6}}|+(-)\frac{5}{2}\rangle + \sqrt{\frac{1}{6}}|-(+)\frac{3}{2}\rangle, \\
|f_2 \Downarrow (\Uparrow)\rangle &= |+(-)\frac{1}{2}\rangle,
\end{aligned}
\tag{5.1}
$$

where $\Downarrow$ ($\Uparrow$) is the pseudo-spin up (down). The kinetic term is given by

$$
\hat{H}_0 = \sum_{k\Sigma} \epsilon_k c^\dagger_{k\Sigma} c_{k\Sigma} + \sum_{kl\sigma} E f^\dagger_{kl\sigma} f_{kl\sigma} + \sum_{kl\sigma\Sigma} \left(V^*_{kl\Sigma\sigma} f^\dagger_{kl\sigma} c_{k\Sigma} + \text{h.c} \right), \tag{5.2}
$$

where $c^\dagger_{k\sigma}$ is a creation operator for s-electron with momentum $\boldsymbol{k}$ and spin σ on Ce-ion. ϵ_k is the conduction band dispersion, which we explain in Appendix F. $f^\dagger_{kl\Sigma}$ is a creation operator for f-electron with $\boldsymbol{k}$, orbital l, pseudo-spin σ, and energy E. $V_{kl\Sigma\sigma}$ is the s-f hybridization term between the nearest Ce sites. Using the Slate-Koster method [39], $V_{kl\Sigma\sigma}$ is given as

$$
V_{kl\Sigma\sigma} = -\sigma t^l_{sf} (\sin k_y + (-1)^l \sigma i \sin k_x) \delta_{\sigma,\Sigma}, \tag{5.3}
$$

where the pseudo-spin is conserved ($\sigma = \Sigma$) in the present 2D model. Detailed explanation of $V_{kl\Sigma\sigma}$ is written in Appendix F. Hereafter, we set $2|t^1_{ss}| = 1$ as energy unit, and put $t^l_{sf} = \sqrt{18/14} \times (0.78)$, $E = -2.0$, $T = 0.01$, and $\mu = -2.45$. Then, $f(s)$-electron number is $n_f = 0.58$ ($n_s = 0.69$). We comment that n_f increases if we put the level of E_l lower under the condition $n_f + n_c$ =const. By this procedure, our main results will not change since the shape of the Fermi surface is essentially unchanged.

Figure 5.1a shows the band structure of the present PAM. The lowest band crosses the Fermi level ($\epsilon = 0$). Since $W_D \sim 5$ eV in CeB_6 [20, 21, 40, 41], $2|t^1_{ss}|(= 1)$

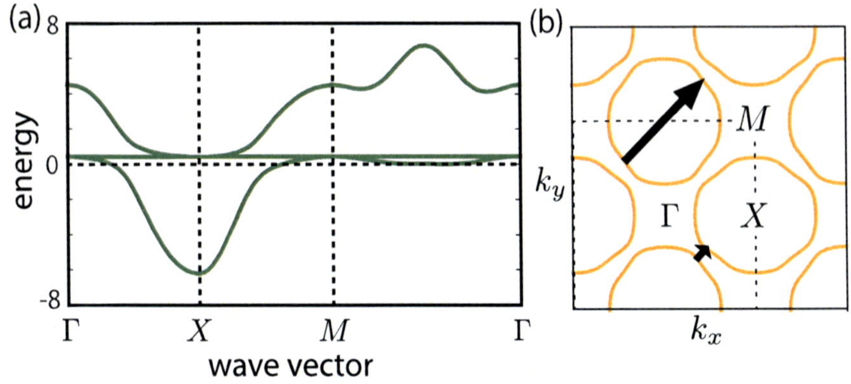

Fig. 5.1 **a** Band dispersion and **b** Fermi surfaces of the present model. Black vectors represent major nesting vectors. Reprinted with permission from Ref. [49]. Copyright ©2019 by the American Physics Society

corresponds to ~0.5eV. The bandwidth of itinerant f-electron is $W_D^{qp} \sim |V_{\mathbf{k}l\Sigma\sigma}| \sim 1$. The Fermi surfaces shown in Fig.4.1b are composed of large ellipsoid electron pockets around X,Y points, consistently with recent ARPES studies [20, 21]. We also consider the Coulomb interaction introduced in Eq. (4.7). The maximum element of $\hat{U}$ is set to unity.

Multipole Symmetry

In Γ_8 quartet model, there are 16-type active multipole operators up to rank 3; monopole, dipole (rank 1), quadrupole (rank 2), and octupole (rank 3) momenta. The table of irreducible representation (IR) in the D_{4h} point group is shown in Table 5.1. Even-rank (odd-rank) operators correspond to electric (magnetic) multipole operators.

Here, we define the Coulomb interaction in the multipole basis, which is given by

$$U^{Q,Q'} = (\mathbf{Q})^{\dagger}\hat{U}\mathbf{Q}'. \tag{5.4}$$

Here, $\mathbf{Q}$ is 16×1 vector Table 5.2 shows the diagonal component $U^Q \equiv U^{Q,Q}$. Electric channels of the Coulomb interaction are much smaller than that for the magnetic channels. Therefore, quadrupole phase cannot be explained by the symmetry of Coulomb interaction. The inner product $(\mathbf{Q})^{\dagger}\mathbf{Q}'$ is unity for $Q = Q'$. It is zero when Q and Q' are different IR, whereas it is not always zero when $Q \neq Q'$ are the same IR [27].

The Green function $\hat{G}^c$, $\hat{G}^f$, $\hat{G}^{cf}$ is defined in the same way as Eqs. (4.21)–(4.22). In the present model, $V^*_{\mathbf{k}l\sigma\sigma}V_{\mathbf{k}m\sigma\sigma}$ in the f-electron Green function $\hat{G}^f$ is given by

$$V^*_{\mathbf{k}l\sigma\sigma}V_{\mathbf{k}m\sigma\sigma} = t^l_{sf}t^m_{sf}(\sin^2 k_y + (-1)^{l+m}\sin^2 k_x + 2(l-m)\sigma i \sin k_x). \tag{5.5}$$

Table 5.1 Irreducible representations and 16-type active multipole operators of D_{4h} point group. Operator with rank k corresponds to 2^k-pole

IR(Γ)	Rank(k)	Multipole($\hat{Q}$)	Matrix($\hat{O}^Q$)	IR in H_z
Γ_1^+	0	$\hat{1}$	$\hat{\sigma}^0\hat{\tau}^0$	Γ_1
	2	$\hat{O}_{20}$	$4\hat{\sigma}^0\hat{\tau}^z$	
Γ_3^+	2	$\hat{O}_{22}$	$4\hat{\sigma}^0\hat{\tau}^x$	Γ_3
Γ_4^+	2	$\hat{O}_{xy}$	$-\hat{\sigma}^z\hat{\tau}^y$	Γ_4
Γ_5^+	2	$\hat{O}_{yz(zx)}$	$-\hat{\sigma}^{x(y)}\hat{\tau}^y$	Γ_5
Γ_2^-	1	$\hat{J}_z$	$\hat{\sigma}^z(-1.2\hat{\tau}^0 - 0.67\hat{\tau}^z$	Γ_1
	3	$\hat{T}_{z\alpha}$	$\hat{\sigma}^z(-\hat{\tau}^0 - 7\hat{\tau}^z)$	
Γ_3^-	3	$\hat{T}_{xyz}$	$-10\hat{\sigma}^0\hat{\tau}^y$	Γ_4
Γ_4^-	3	$\hat{T}_{z\beta}$	$-6.7\hat{\sigma}^z\hat{\tau}^x$	Γ_3
Γ_5^-	1	$\hat{J}_{x(y)}$	$\hat{\sigma}^{x(y)}(1.2\hat{\tau}^0 - 0.34\hat{\tau}^z + (-)0.58\hat{\tau}^x)$	Γ_5
	3	$\hat{T}_{x\alpha(y\alpha)}$	$\hat{\sigma}^{x(y)}(\hat{\tau}^0 - (+)3.5\hat{\tau}^z + 6.1\hat{\tau}^x)$	
	3	$\hat{T}_{x\beta(y\beta)}$	$\hat{\sigma}^{x(y)}(-5.8\hat{\tau}^z - (+)3.4\hat{\tau}^x)$	

Table 5.2 Normalized Coulomb interaction U^Q. The relation $U^{Q,Q'} = 0$ holds for $Q \neq Q'$ except for $U^{J_\mu, T_\mu^\alpha} = 0.58$ where $\mu = x, y, z$. Reprinted with permission from Ref. [49]. Copyright ©2019 by the American Physics Society

Q	1	$O_{20(22)}$	$O_{xy(yz,zx)}$	T_{xyz}	$J_{z(x,y)}$	$T^\alpha_{z(x,y)}$	$T^\beta_{z(x,y)}$
U^Q	-2.4	0.50	0.63	0.81	1.03	0.94	0.94

Therefore, the symmetry of the Green function is given by

$$\hat{G}^f \propto \hat{\sigma}^0(\hat{\tau}^0 + a\hat{\tau}^x) + b\hat{\sigma}^z\hat{\tau}^y. \tag{5.6}$$

The f-electron Green function has spin dependence,

$$\begin{aligned} G^f_{ll}(k) &\equiv G^f_{ll\uparrow}(k) = G^f_{l\downarrow}(k) \quad (G^f_{lm\uparrow}(k) \neq G^f_{lm\downarrow}(k)), \\ G^c(k) &\equiv G^c_\uparrow(k) = G^c_\downarrow(k), \end{aligned} \tag{5.7}$$

which means that the spin index cannot be excluded unlike Γ_7 model.

- We consider 2D Γ_8 quartet PAM in the presence of strong SOI.

- We cannot explain O_{xy} quadrupole phase by RPA since electric ch of bare Coulomb interaction is much smaller than magnetic ch ones.
- f-electron Green function is $\hat{G}^f \propto \hat{\sigma}^0(\hat{\tau}^0 + a\hat{\tau}^x) + b\hat{\sigma}^z\hat{\tau}^y$.

5.3 Results

5.3.1 Magnetic Multipole Susceptibility

The bare irreducible susceptibility is given by

$$\chi^0_{\alpha,\beta}(q) = -T\sum_k G^f_{LM}(k+q)G^f_{M'L'}(k), \tag{5.8}$$

where $L \equiv (l, \sigma)$ and $\alpha \equiv (L, L')$. α, β takes $1 \sim 16$, and $\hat{G}^f$. Here we use the Green function without self-energy. Then, f-electron susceptibility within RPA is given as

$$\hat{\chi}(q) = \hat{\chi}^0(q)(\hat{1} - u\hat{U}\hat{\chi}^0(q))^{-1}. \tag{5.9}$$

Here, we consider the following eigen equation:

$$u\hat{U}\hat{\chi}^0(\boldsymbol{q}, 0)\mathbf{w}^\Gamma(\boldsymbol{q}) = \alpha^\Gamma(\boldsymbol{q})\mathbf{w}^\Gamma(\boldsymbol{q}), \quad (\mathbf{w}^\Gamma(\boldsymbol{q}) = \sum_{Q\in\Gamma} Z^Q(\boldsymbol{q})\mathbf{Q}), \tag{5.10}$$

where $Z^Q(\boldsymbol{q})$ is a real coefficient. The maximum of the eigenvalue $\alpha^\Gamma(\boldsymbol{q})$ gives the Stoner factor for IR Γ,

$$\alpha^\Gamma = \max_{\boldsymbol{q}}\{\alpha^\Gamma(\boldsymbol{q})\}, \qquad / \qquad \alpha^{\mathrm{mag(el)}} = \max_n\{\alpha^{\Gamma_n^{-(+)}}\}, \quad (n = 1, 3, 5) \tag{5.11}$$

where $\alpha^{\mathrm{mag(el)}}$ is the magnetic (electric) Stoner factor. The Γ-channel multipole order appears when $\alpha^\Gamma \geq 1$. Using $\mathbf{Q}$, the multipole susceptibility is given by

$$\chi^{Q,Q'}(q) = (\mathbf{Q})^\dagger\hat{\chi}(q)\mathbf{Q}'. \tag{5.12}$$

First, we show the numerical results by RPA. Figure 5.2 shows obtained susceptibilities $\chi^Q(\boldsymbol{q}, 0) \equiv \chi^{Q,Q}(\boldsymbol{q}, 0)$ at $u = 1.08$ ($\alpha^{\mathrm{mag}} = 0.9$). In RPA, χ^{J_z} is the most largest. Secondly, $\chi^{T^\beta_\nu}$, $\chi^{T^\alpha_\nu}$ ($\nu = x, y$), and $\chi^{T_{xyz}}$ are also enlarged. $\chi^{J_z}(\boldsymbol{q}, 0)$ has peak value at $\boldsymbol{q} \approx \mathbf{0}$ and $\boldsymbol{q} \approx \boldsymbol{Q} \equiv (\pi, \pi)$, which is consistent with the inelastic neutron scattering that reports strong ferromagnetic and anti-ferromagnetic fluctuations at

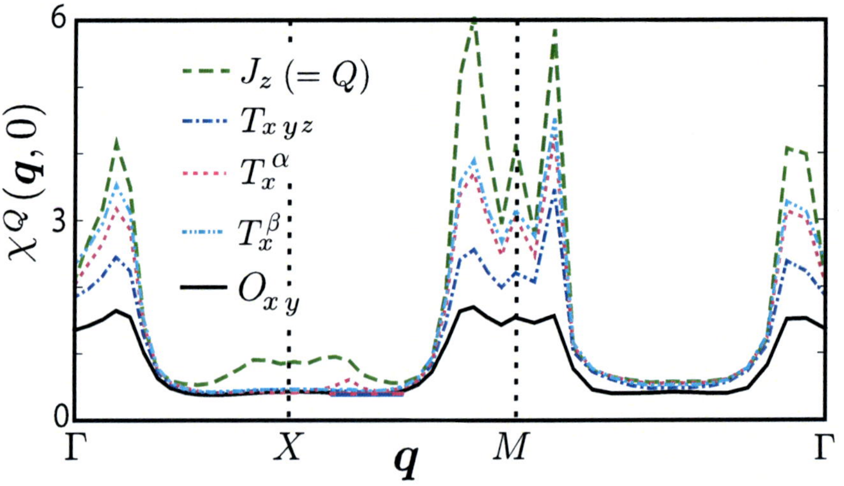

Fig. 5.2 Obtained multipole susceptibilities by RPA. The peak positions correspond to the nesting vectors in Fig. 5.1b. Reprinted with permission from Ref. [49]. Copyright ©2019 by the American Physics Society

$\boldsymbol{q} = (\pi, \pi, \pi), (\pi, \pi, 0)$ above T_N [23, 42]. Therefore, the present two-dimensional PAM is reliable. On the other hand, RPA quadrupole susceptibility remains small.

- Various magnetic fluctuations develop within RPA.
- Obtained magnetic susceptibilities have peak at $\boldsymbol{q} \approx \boldsymbol{0}$ and $\boldsymbol{q} \approx \boldsymbol{Q} \equiv (\pi, \pi)$.

5.3.2 *Electric Multipole Susceptibility*

From now on, we introduce the VC due to AL- and MT-terms.

$$\hat{\chi}(q) = \hat{\phi}(q)(\hat{1} - u\hat{U}\hat{\phi}(q))^{-1}, \quad (\hat{\phi}(q) \equiv \hat{\chi}^0(q) + \hat{X}^{\mathrm{AL+MT}}(q)), \tag{5.13}$$

where $\hat{\phi}(q)$ is irreducible susceptibility including the VC in the 16×16 matrix form. Diagrams of these VC are shown in Fig. 5.3. For example, the expression for the MT-term is

$$X^{MT}_{LL'MM'}(q) = T^2 \sum_{p,k,A\sim D} G_{LA}(k+q-p)G_{BL'}(k-p)G_{DM}(k+q)G_{M'C}(k)V_{DACB}(p), \tag{5.14}$$

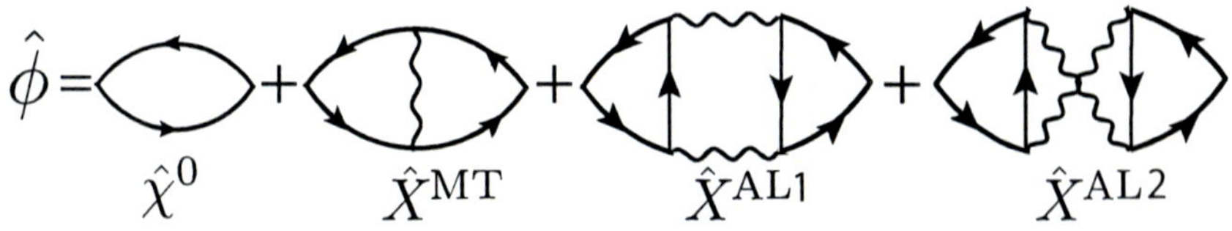

Fig. 5.3 Diagrams of the irreducible susceptibility $\hat{\phi}$ with MT- and AL-VC

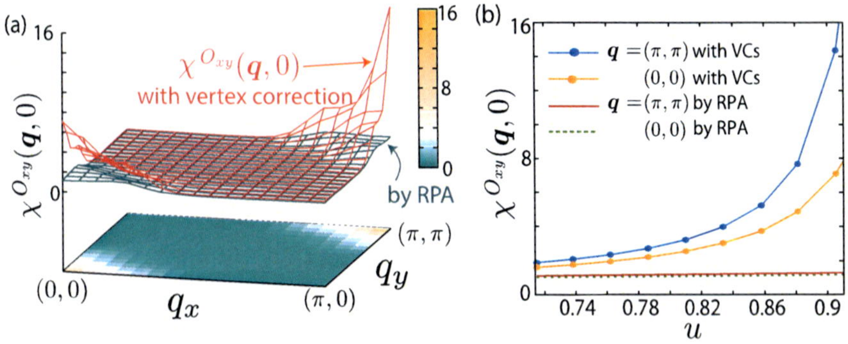

Fig. 5.4 **a** $\boldsymbol{q}$-dependence of $\chi^{O_{xy}}(\boldsymbol{q},0)$ at $\alpha^{\Gamma_4^+}=0.94$ with AL-VC+MT-VC. **b** u-dependence of $\chi^{O_{xy}}(\boldsymbol{q},0)$ at $\boldsymbol{q}=\boldsymbol{Q}$ and $\boldsymbol{q}=\boldsymbol{0}$. Reprinted with permission from Ref. [49]. Copyright ©2019 by the American Physics Society

where $p \equiv (\boldsymbol{p}, \omega_m)$, and $\hat{V}(q) \equiv u^2\hat{U}\hat{\chi}(q)\hat{U} + u\hat{U}$ is the dressed interaction given by RPA.

Figure 5.4a and b shows the obtained quadrupole susceptibility including MT- and AL-VC. In contrast to RPA result, the obtained $\chi^{O_{xy}}(\boldsymbol{q},0)$ is strongly enhanced at $\boldsymbol{q}=\boldsymbol{Q}$ and $\boldsymbol{q}=\boldsymbol{0}$, and becomes the largest of all χ^Q. This enhancement originates from the AL-terms, whereas the MT-term is very small as we have shown in Fig. 5.5. We find that $\chi_{\rm AL}^{O_{xy}}(\boldsymbol{q},0)$ strongly increases with u. In contrast, $\chi_{\rm MT}^{O_{xy}}(\boldsymbol{q},0)$ remains small and comparable to RPA result in Fig. 5.4. Therefore, it is verified that the enhancement of O_{xy} quadrupole fluctuations originates from the AL-VC, whereas the MT-VC is very small.

The obtained $\chi^{O_{xy}}(\boldsymbol{q},0)$ has the highest peak at $\boldsymbol{q}=\boldsymbol{Q}$, consistently with the anti-ferro-O_{xy} order in CeB_6. Moreover, the second highest peak of $\chi^{O_{xy}}(\boldsymbol{q},0)$ at $\boldsymbol{q}=\boldsymbol{0}$ explains the softening of shear modulus C_{44} in CeB_6 [10].

In Fig. 5.5, we show all quadrupole susceptibilities including O_{xy}, O_{zx}, O_{yz}, O_{20}, O_{22}. In the present 2D model, only O_{xy}-fluctuation strongly develops. The reason is that (T_x, T_y) fluctuations are much larger than T_z fluctuations in RPA due to the violation of cubic symmetry. Since $O_{\mu\nu}$ quadrupole susceptibility is magnified by (T_μ, T_ν) fluctuations ($\mu, \nu = x, y, z$) due to the AL-VC, $\chi^{O_{xy}}(\boldsymbol{q},0)$ is larger than that for O_{zx}, O_{yz} in the present 2D model. In contrast, in the cubic model, $\chi^Q(\boldsymbol{q},0)$ with $Q = O_{xy}, O_{zx}, O_{yz}$ should equally develop.

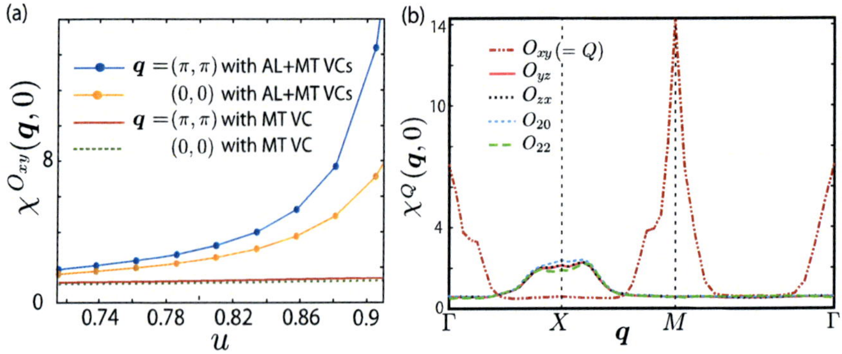

Fig. 5.5 **a** Obtained $\chi_{\mathrm{AL}}^{O_{xy}}(\boldsymbol{q}, 0)$ with AL1+AL2-terms and $\chi_{\mathrm{MT}}^{O_{xy}}(\boldsymbol{q}, 0)$ with MT-term at $\boldsymbol{q} = \boldsymbol{Q}, \boldsymbol{0}$ as function of u. **b** Obtained quadrupole susceptibilities $\chi^{Q}(\boldsymbol{q}, 0)$ for $Q = O_{xy}, O_{zx/yz}$, and $O_{20/22}$. Reprinted with permission from Ref. [49]. Copyright © 2019 by the American Physics Society

In addition, as we have shown in Table 5.2, the Coulomb interaction U^{Q} for $Q = O_{xy/yz/zx}$ is much larger than that for $Q = O_{20/22}$. For this reason, it is difficult to expect that $Q = O_{20/22}$ quadrupole susceptibility becomes larger than $Q = O_{xy}$ one, even if the AL-VC are considered. Thus, the relation $\chi^{O_{xy}}(\boldsymbol{q}, 0) > \chi^{O_{20/22}}(\boldsymbol{q}, 0)$ should hold even in cubic systems. To summarize, the obtained strong enhancements of $\chi^{O_{xy}}(\boldsymbol{q}, 0)$ and $\chi^{J_z}(\boldsymbol{q}, 0)$ at both $\boldsymbol{q} = \boldsymbol{Q}$ and $\boldsymbol{q} = \boldsymbol{0}$ reproduce the key experimental results of CeB_6.

- $\chi^{O_{xy}}(\boldsymbol{q}, 0)$ develops by considering AL- and MT-terms at both $\boldsymbol{q} = \boldsymbol{Q}$ and $\boldsymbol{q} = \boldsymbol{0}$.
- AL-term brings dominant contribution to the enhancement of $\chi^{O_{xy}}$.
- $\chi^{O_{yz(zx)}}$ should equally develop if we consider 3D nature of CeB_6.

5.4 Discussion

5.4.1 Microscopic Origin of Quadrupole Phase

Next, we explain that the O_{xy} quadrupole order is derived from the interference between magnetic multipole fluctuations. For this purpose, we analyze the total AL-term for O_{xy}-channel defined as

$$X_{O_{xy}}^{\mathrm{AL}}(q) \equiv (\mathbf{O}_{xy})^{\dagger} \hat{X}^{\mathrm{AL}}(q) \mathbf{O}_{xy}. \tag{5.15}$$

The Stoner factor for $\chi^{O_{xy}}$ channel is proportional to $uU^{O_{xy}}\phi^{O_{xy}}(q)$, where $\phi^{O_{xy}}(q) \equiv (\mathbf{O}_{xy})^\dagger \hat{\phi}(q)\mathbf{O}_{xy}$. Therefore, $X^{\mathrm{AL}}_{O_{xy}}(q)$ (> 0) works as enhancement factor of O_{xy} susceptibility. For instance, the AL1-term due to (Q, Q')-channel fluctuations is defined as

$$X^{\mathrm{AL1},QQ'}_{O_{xy}}(q) \equiv \frac{T}{2}\sum_p V^Q(p)V^{Q'}(p-q)\Lambda^{O_{xy}QQ'}_{q,p}(\Lambda^{O_{xy}Q'Q}_{\bar{q},\bar{p}})^*, \tag{5.16}$$

where $V^Q(q)$ and $\Lambda^{O_{xy}QQ'}_{q,p}$ are defined as

$$\hat{V}(q) = \sum_Q V^Q(q)\mathbf{Q}(\mathbf{Q})^\dagger, \tag{5.17}$$

$$\Lambda^{O_{xy}QQ'}_{q,p} \equiv \sum_\alpha (\mathbf{O}_{xy})^*_\alpha (\mathbf{Q}')^\dagger \hat{\Lambda}^\alpha(q,p)\mathbf{Q}. \tag{5.18}$$

The diagrammatic expression of Eq. (5.16) is shown in Fig. 5.6a. Figure 5.6b shows the $\boldsymbol{q}$-dependence of $X^{QQ'}_{O_{xy}}(\boldsymbol{q}, 0)$ at $u = 0.91$. We find that the $(Q, Q') = (T^\alpha_x, T^\alpha_y)$, (J_z, T_{xyz}), (T^β_x, T^β_y) channels give the dominant contributions. Other terms not shown in Fig. 5.6b give negligible contribution. Therefore, we conclude that O_{xy} quadrupole fluctuation is enhanced by various types of magnetic fluctuations.

Next, we discuss the $\boldsymbol{q}$-dependence of the AL-VC, which is given as

$$X^{AL,T_xT_y}_{O_{xy}}(\boldsymbol{q}) \propto \sum_p \chi^{T_x}(\boldsymbol{p})\chi^{T_y}(\boldsymbol{q}-\boldsymbol{p}). \tag{5.19}$$

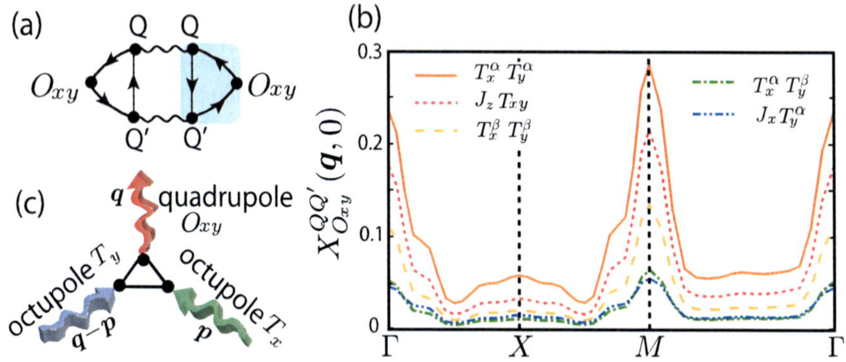

Fig. 5.6 **a** AL-term $X^{\mathrm{AL1},QQ'}_{O_{xy}}$ given by (Q, Q')-channel fluctuations. **b** Obtained $X^{QQ'}_{O_{xy}}(\boldsymbol{q}, 0)$ along high-symmetry line. **c** Quantum process of O_{xy} fluctuations driven by the interference between (T_x, T_y) fluctuations, which corresponds to the shaded area in **a**. Note that $\chi^{O_{xy}} \neq \chi^{O_{xz(yz)}}$ in the present 2D model. Reprinted with permission from Ref. [49]. Copyright © 2019 by the American Physics Society

It becomes large at $\boldsymbol{q} = \boldsymbol{Q}$ and $\boldsymbol{q} = \boldsymbol{0}$ since $\chi^{T_\mu}(\boldsymbol{p})$ has large peaks at $\boldsymbol{p} \sim \boldsymbol{Q}, \boldsymbol{0}$ as shown in Fig. 5.2. Thus, anti-ferroquadrupole order in CeB_6 originates from the interference between ferro- and anti-ferromagnetic multipole fluctuations.

Figure 5.6c presents the quantum process of $\hat{O}_{xy}$ quadrupole order driven by the interference between (T_x, T_y) fluctuations, which corresponds to $\Lambda^{O_{zx}T_xT_y}$ in Fig. 5.6a. This process is realized when

$$\Lambda^{O_{zx}QQ'} \propto \mathrm{Tr}\{\hat{O}_{xy} \cdot \hat{Q} \cdot \hat{Q}'\} \neq 0. \tag{5.20}$$

In contrast, since $\Lambda^{QT_xT_y} \propto \mathrm{Tr}\{\hat{Q} \cdot \hat{T}_x \cdot \hat{T}_y\} = 0$ for odd-rank Q, the AL-VC is negligible for $Q = J, T$.

- Microscopic origin of $\chi^{O_{xy}}$ is interference between two types of magnetic fluctuations, mainly from T_x and T_y fluctuations.
- The peak structure of $\chi^{O_{xy}}(\boldsymbol{q})$ originates from large $\chi^{T_x,T_y}(\boldsymbol{p})$ at $\boldsymbol{p} \sim \boldsymbol{Q}, \boldsymbol{0}$.

5.4.2 Magnetic Field

Finally, we discuss the field-induced octupole order, which has been studied intensively as a main issue of CeB_6 [13–16]. The Zeeman term under the magnetic field along z-axis is given as

$$\hat{H}_Z = h_z \sum_{L,M} (\hat{J}_z)_{L,M} f^\dagger_{\boldsymbol{k}L} f_{\boldsymbol{k}M}. \tag{5.21}$$

In the presence of magnetic field $h_z \neq 0$, electric and magnetic multipole belongs to the same IR as shown in the fifth column in Table 5.1 [13]. In this case, both O_{xy} and T_{xyz} belong to the same IR Γ_4 Therefore, inter-multipole susceptibility $\chi^{O_{xy},T_{xyz}}(\boldsymbol{q}, 0)$ is induced in proportion to h_z, since

$$\mathrm{Tr}\{\hat{O}_{xy} \cdot \hat{J}_z \cdot \hat{T}_{xyz}\} \neq 0. \tag{5.22}$$

In Fig. 5.7a, we show the diagrammatic expression of quadrupole-octupole susceptibility in the presence of magnetic field. To verify this, we solve the eigen equation (5.10) for the IR Γ_4 under h_z, at the fixed magnetic Stoner factor in RPA at $\alpha^{\mathrm{mag}} = 0.8$ [43, 44].

Figure 5.7b shows the obtained coefficient of eigenvector $\mathbf{w}^{\Gamma_4}$, which is defined as

$$\mathbf{w}^{\Gamma_4}(\boldsymbol{q}) = Z^{O_{xy}}(\boldsymbol{q})\mathbf{O}_{xy} + Z^{T_{xyz}}(\boldsymbol{q})\mathbf{T}_{xyz} \quad (|\mathbf{w}^{\Gamma_4}|^2 = 1). \tag{5.23}$$

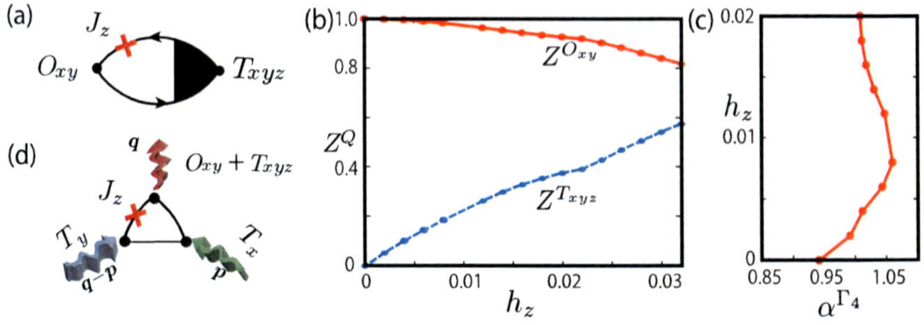

Fig. 5.7 **a** h_z-linear term of the three-point vertex $\Lambda^{T_{xyz}T_xT_y}$ that gives large $\chi^{O_{xy}T_{xyz}}(\boldsymbol{q}, 0)$. **b** Form factor $(Z^{O_{xy}}, Z^{T_{xyz}})$ of the eigenvector for $\Gamma_4 = \{O_{xy}, T_{xyz}$ at $\boldsymbol{q} = \boldsymbol{Q}$ under h_z. **c** Stoner factor α^{Γ_4} as function of h_z. **d** Quantum interference mechanism among the magnetic multipole fluctuations. Reprinted with permission from Ref. [49]. Copyright ©2019 by the American Physics Society

$Z^{T_{xyz}}$ increases linearly in h_z, due to the interference process under h_z shown in Fig. 5.7a. $Z^{T_{xyz}}$ becomes comparable to $Z^{O_{xy}}$ under small magnetic field $h_z \lesssim 0.03 \ll W_D^{qp}/10$.

Figure 5.7c shows the largest Stoner factor α^{Γ_4} at $\boldsymbol{q} = \boldsymbol{Q}$ as functions of h_z. The increment of α^{Γ_4} under h_z is consistent with the field enhancement of T_Q in CeB_6. (In contrast, T_N will be suppressed by large O_{xz} moment.) Since the ratio of the ordered momenta at T_Q is $M^{T_{xyz}}/M^{O_{xy}} = Z^{T_{xyz}}/Z^{O_{xy}}$, field-induced anti-ferro T_{xyz} order is naturally explained. In Fig. 5.7(d), we show the proposed mechanism of field-induced $O_{xy} + T_{xyz} \in \Gamma_4$ order. h_z-linear term – large $(Z^{O_{xy}}, Z^{T_{xyz}})$.

- Anti-ferro T_{xyz} phase is induced by magnetic field along z-axis since $\chi^{O_{xy},T_{xyz}} \neq 0$.
- T_{xyz} order is strongly magnified by large $\chi^{O_{xy},T_{xyz}}$, which comes from AL-type χ-VC.
- Origin of T_{xyz} phase is interference of multipole fluctuations.

Conclusion

In summary, we proposed multipole fluctuation mechanism to explain the quadrupole ordering in CeB_6 by considering AL-VC in HF systems. As a result, both ferro- and anti-ferromagnetic multipole fluctuations are induced around nesting vector of the Fermi surface. It is consistent with recent neutron experiments. In particular, we showed that AF-O_{xy} order at T_Q ($> T_N$) originates from the interference among the different types of magnetic multipole fluctuations, which are enlarged within RPA as shown in Fig. 5.6c. Also, we verified that magnetic field induces the octupole-ordered phase identified in CeB_6. The inter-multipole coupling mechanism will be important even in other HF systems [45, 46] as well as $4d$, $5d$ transition metal system [47]. Therefore, it is an important future problem to analyze AL-VC in 3D system.

At the starting point of the present study, we showed that on-site quadrupole (O_{xy}) interaction is about 60% of dipole (J_μ) one as listed in Table 5.2. For this reason, quadrupole-ordered phase cannot appear within RPA scheme. In contrast, by using RKKY model, quadrupole interaction gets as large as the dipole one [13, 16, 48]. Therefore, the difference between itinerant and localized scheme may enable us to notice some important facts.

Finally, the main results of the present study are summarized as follows:

1. Near the AFM-QCP, several multipole fluctuations strongly develop, simultaneously including higher rank (octupole T) fluctuations.
2. Development of magnetic multiple multipole fluctuations gives large χ-VC for electric ch fluctuations, which cause violation of RPA.
3. Owning to χ-VC, AF-quadrupole fluctuation $\chi^{O_{xy}}(\boldsymbol{q})$ at $\boldsymbol{q} = (\pi, \pi)$ develops due to the interference between magnetic octupole fluctuations given by T_x and T_y at $\boldsymbol{p} \sim \boldsymbol{Q}, \boldsymbol{0}$.
4. Anti-ferro T_{xyz} phase is induced by small magnetic field along z-axis due to the χ-VC and inter-multipole fluctuation $\chi^{O_{xy}, T_{xyz}}$.

Appendix 4

Here, we explain the model Hamiltonian for CeB_6. The conduction band is $5d$ electrons on Ce-ions. In the present study, to simplify the model, we introduce conduction band composed of s electrons. The effective tight-binding parameters of CeB_6 are given in Ref. [21] in Chap. 5. We use slightly modified model and put $k_z = 0$, in order to reproduce the experimental Fermi surfaces of CeB_6 on the k_x-k_y-plane after s-f hybridization, which is given by

$$\begin{aligned}\epsilon_{\boldsymbol{k}} = {}& t_{ss}^1\left(\cos k_x + \cos k_y\right) + t_{ss}^2\left\{\cos(k_x + k_y) + \cos(k_x - k_y)\right\} + t_{ss}^3\left(\cos 2k_x + \cos 2k_y\right)\\ &+ t_{ss}^4\left\{\cos(2k_x + k_y) + \cos(2k_x - k_y) + \cos(2k_y + k_x) + \cos(2k_y - k_x)\right\}\\ &+ t_{ss}^5\left\{\cos(2k_x + 2k_y) + \cos(2k_x - 2k_y)\right\} + E_0,\end{aligned} \tag{5.24}$$

where t_{ss}^i is the i-th nearest s-s hopping integral. We set $(t_{ss}^1, t_{ss}^2, t_{ss}^3, t_{ss}^4, t_{ss}^5) = (-0.5, -0.889, 0.292, -0.229, 0.687)$, and $E_0 = 1.33$.

Next, we explain the hybridization term. Based on the Slater-Koster tight-binding method, the s-f hybridization between the nearest Ce sites is

$$\begin{aligned}V_{\boldsymbol{k} f_1 \uparrow} &= -A_1 t_{sf}(\sin k_y - i \sin k_x),\\ V_{\boldsymbol{k} f_2 \uparrow} &= -A_2 t_{sf}(\sin k_y + i \sin k_x),\end{aligned} \tag{5.25}$$

and $V_{\boldsymbol{k} f_l \downarrow} = -V^*_{\boldsymbol{k} f_l \uparrow}$. and $A_1 = \sqrt{18/14}$ and $A_2 = \sqrt{3/7}$. Since $A_1 > A_2$, the relation $D_{f_1}(0) > D_{f_2}(0)$ holds in the present two-dimensional PAM, where $D_{f_l}(0)$ is the f_l-electron density of states at Fermi level. However, $D_{f_1}(0) = D_{f_2}(0)$ holds in the cubic model, since the s-f hybridization along z-axis is larger for f_2-electron. To escape from the artifact of two dimensionality, we put $A_1 = A_2 = \sqrt{18/14}$ in the present study. In the present Γ_8 model, the relation $V_{\boldsymbol{k} f_1 \sigma} \propto V^*_{\boldsymbol{k} f_2 \sigma}$ holds. In contrast, in the $\Gamma_7^{(1)}$-$\Gamma_7^{(2)}$ model for $CeCu_2Si_2$ used in Appendix C, the relation $V_{\boldsymbol{k} f_1 \sigma} \propto V_{\boldsymbol{k} f_2 \sigma}$ holds.

References

1. P. Coleman, *Handbook of Magnetism and Advanced Magnetic Materials*, vol. 1 (Wiley, 2007), pp. 95–148
2. T. Moriya, K. Ueda, Adv. Phys. **49**, 555 (2000)
3. Y. Yanase, T. Jujo, T. Nomura, H. Ikeda, T. Hotta, K. Yamada, Phys. Re. **387**, 1 (2003)
4. H. Kontani, Rep. Prog. Phys. **71** (2008)
5. P. Monthoux, D. Pines, G.G. Lonzarich, Nature **450**, 1177 (2007)
6. D. Senechal, A.-M.S. Tremblay, Phys. Rev. Lett. **92** (2004)
7. P. Monthoux, D.J. Scalapino, Phys. Rev. Lett. **72**, 1874 (1994)
8. T. Takimoto, T. Hotta, K. Ueda, J. Phys.: Condens. Matter **15**, S2087 (2003)
9. T. Fujita, M. Suzuki, T. Komatsubara, S. Kunii, T. Kasuya, T. Ohtsuka, Solid State Commun. **35**, 569 (1980)
10. S. Nakamura, T. Goto, S. Kunii, K. Iwashita, A. Tamaki, J. Phys. Soc. Jpn. **63**, 623 (1994)
11. M. Hiroi, S. Kobayashi, M. Sera, N. Kobayashi, S. Kunii, J. Phys. Soc. Jpn. **66**, 132 (1997)
12. A.S. Cameron, G. Friemel, D.S. Inosov, Rep. Prog. Phys. **79** (2016)
13. R. Shiina, S. Shiba, P. Thalmeier, J. Phys. Soc. Jpn. **66**, 1741 (1997)
14. O. Sakai, R. Shiina, H. Shiba, P. Thalmeier, J. Phys. Soc. Jpn. **66**, 3005 (1997)(mohana); O. Sakai, R. Shiina, H. Shiba, P. Thalmeier, J. Phys. Soc. Jpn. **68**, 1364 (1999)
15. P. Thalmeier, R. Shiina, H. Shiba, O. Sakai1, J. Phys. Soc. Jpn. **67**, 2363 (1998)
16. H. Shiba, O. Sakai1, R. Shiina, J. Phys. Soc. Jpn. **68**, 1988 (1999)
17. M. Sera, S. Kobayashi, J. Phys. Soc. Jpn. **68**, 1664 (1999)
18. H. Kusunose, Y. Kuramoto, J. Phys. Soc. Jpn. **74**, 3139 (2005)
19. K. Hanzawa, J. Phys. Soc. Jpn. **70**, 468 (2001)
20. M. Neupane, N. Alidoust, I. Belopolski, G. Bian, S.-Y. Xu, D.-J. Kim, P.P. Shibayev, D.S. Sanchez, H. Zheng, T.-R. Chang, H.-T. Jeng, P.S. Riseborough, H. Lin, A. Bansil, T. Durakiewicz, Z. Fisk, M.Z. Hasan, Phys. Rev. B **92** (2015)
21. A. Koitzsch, N. Heming, M. Knupfer, B. Buchner, P.Y. Portnichenko, A.V. Dukhnenko, N.Y. Shitsevalova, V.B. Filipov, L.L. Lev, V.N. Strocov, J. Ollivier, D.S. Inosov, Nat. Commun. **7**, 10876 (2016)
22. G. Friemel, Y. Li, A.V. Dukhnenko, N.Y. Shitsevalova, N.E. Sluchanko, A. Ivanov, V.B. Filipov, B. Keimer, D.S. Inosov, Nat. Commun. **3**, 830 (2012)
23. H. Jang, G. Friemel, J. Ollivier, A.V. Dukhnenko, NYu. Shitsevalova, V.B. Filipov, B. Keimer, D.S. Inosov, Nat. Mater. **13**, 682 (2014)
24. A. Akbari, P. Thalmeier, Phys. Rev. Lett. **108** (2012)
25. M. Endo, S. Nakamura, T. Isshiki, N. Kimura, T. Nojima, H. Aoki, H. Harima, S. Kunii, J. Phys. Soc. Jpn. **75** (2006)
26. H. Ikeda, M.-T. Suzuki, R. Arita, T. Takimoto, T. Shibauchi, Y. Matsuda, Nat. Phys. **8**, 528 (2012)
27. R. Tazai, H. Kontani, Phys. Rev. B **98** (2018)

28. S. Onari, H. Kontani, Phys. Rev. Lett. **109** (2012)
29. S. Onari, Y. Yamakawa, H. Kontani, Phys. Rev. Lett. **116** (2016)
30. Y. Yamakawa, S. Onari, H. Kontani, Phys. Rev. X **6** (2016)
31. M. Tsuchiizu, Y. Ohno, S. Onari, H. Kontani, Phys. Rev. Lett. **111**, 057003 (2013)
32. M. Tsuchiizu, Y. Yamakawa, S. Onari, Y. Ohno, H. Kontani, Phys. Rev. B **91** (2015)
33. M. Tsuchiizu, Y. Yamakawa, H. Kontani, Phys. Rev. B **93** (2016)
34. R. Tazai, Y. Yamakawa, M. Tsuchiizu, H. Kontani, Phys. Rev. B **94** (2016)
35. M. Tsuchiizu, K. Kawaguchi, Y. Yamakawa, H. Kontani, Phys. Rev. B **97** (2018)
36. R.-Q. Xing, L. Classen, A.V. Chubukov, Phys. Rev. B **98**, 041108 (2018)
37. U. Karahasanovic, F. Kretzschmar, T. Bohm, R. Hackl, I. Paul, Y. Gallais, J. Schmalian, Phys. Rev. B **92** (2015)
38. R. Tazai, Y. Yamakawa, M. Tsuchiizu, H. Kontani, J. Phys. Soc. Jpn. **86** (2017)
39. K. Takegahara, Y. Aoki, A. Yanase, J. Phys. C **13**, 583 (1980)
40. Y. Kubo, S. Asano, Phys. Rev. B **39**, 8822 (1989)
41. A. Hasegawa, A. Yanase, J. Phys. F. Metal Phys. **7**, 7 (1977)
42. At $T \sim T_Q$, $\chi^{J_\mu}(\boldsymbol{q}, \omega)$ at $= \boldsymbol{Q}, \boldsymbol{0}$ is much larger than that at the magnetic order wavevector $\boldsymbol{Q}_N = (\pi/2, \pi/2, 0)$ below T_N according to Ref. [23]
43. We adjust u to keep α^{mag} constant, since α^{mag} decreases with h_z for fixed u in RPA. In the FLEX approximation, α^{mag} increases with h_z due to the negative feedback effect between spin fluctuations and self-energy [45]
44. K. Sakurazawa, H. Kontani, T. Saso, J. Phys. Soc. Jpn. **74**, 271 (2005)
45. T. Takimoto, P. Thalmeier, Phys. Rev. B **77** (2008)
46. K. Kubo, T. Hotta, Phys. Rev. B **71**, 140404(R) (2005)
47. S. Hayami, Y. Yanagi, H. Kusunose, Y. Motome, Phys. Rev. Lett. **122** (2019)
48. T. Yamada, K. Hanzawa, J. Phys. Soc. Jpn. **88** (2019)
49. R. Tazai, H. Kontani, Phys. Rev. B **100**, 241103 (R) (2019)

Chapter 6
Summary of This Book

In the present study, we studied microscopic origin of SC and multipole phase in SCES in a unified way considering many body effects beyond RPA+ME theory. Until recently, it was widely believed that the basic properties of SC states in SCES are qualitatively well understood within conventional RPA+ME theory. However, recent experiments revealed that RPA+ME theory fails in several SCES, such as Fe-based and HF superconductor. Thus, new microscopic theory is required to solve the problem. Here, we focused on significant roles of VC due to many body effects. It was revealed that violation of RPA+ME theory occurs especially in multi-orbital and multipole systems. As we have shown in this paper, VC is classified into χ-VC and U-VC;

- VC for p-h irreducible susceptibility (χ-VC), which is absent in RPA.
- VC for SC paring interaction (U-VC), which is neglected in the ME.

We conclude that VC is necessary to understand fundamental properties of various phenomena observed in SCES. Our main findings based on the beyond RPA+ME theory are listed as follows:

1. Both U-VC and χ-VC work as a mode coupling between different types of p-h fluctuations, as we verified by using the fRG and perturbation theory.
2. Mode coupling effects drastically change the ground state. For instance, orbital-ordered state is induced by interference between different magnetic fluctuations near AFM-QCP.
3. U-VC strongly enhances the attractive SC paring interaction due to the electric- or orbital-channel fluctuations.
4. AFM fluctuations and el-ph interaction can work cooperatively on s-wave SC paring mechanism against common knowledge. This result originates from the fact that phonon-mediated attraction is enhanced by U-VC near AFM-QCP.

R. Tazai, *Theory of Multipole Fluctuation Mediated Superconductivity and Multipole Phase*, Springer Theses,
https://doi.org/10.1007/978-981-16-1026-4_6

5. Mode coupling becomes more significant in f-electron system than d-electrons due to large number of multipole degrees of freedom. Higher rank electric ch multipole fluctuations stabilize s-wave SC in $CeCu_2Si_2$.
6. Interference between two types of octupole fluctuations induce quadrupole phase in CeB_6.

We believe that our findings are meaningful step for revealing essential roles of many body effects behind various long-standing problems in SCES. Also, a lot of interesting issues are left in the field of SCES. One of that is nontrivial transport phenomena, such as non-Fermi-liquid behavior, which are reported near the SC and multipole phase. In addition, external-field-induced phases were also discovered. Therefore, we need to continue our work to solve these issues in the future. Hereafter, we review each section (Chaps. 2–5) in more detail.

Functional Renormalization Group (fRG) Study

In Chap. 2, we studied two-orbital (d_{xz}, d_{yz}) Hubbard model without SOI based on fRG+cRPA theory. The fRG study has great merit for understanding fundamental properties of many body effect since parquet-type higher order processes are automatically generated in unbiased way. In fRG+cRPA study, we consider three energy regions with lower cutoff energy ω_c and boundary of fRG and cRPA Λ_0. Hereafter, we set $\Lambda_0 \ll \epsilon_F$. In the highest energy region ($\Lambda_0 < \xi_k$), p-h instability within RPA process is the most dominant. In this case, only spin fluctuations develop while orbital ones remain small. On the other hand, in the lower energy region ($\omega_c < \xi_k < \Lambda_0$), AL-type VC for p-h fluctuation becomes significant and orbital fluctuations develop due to the χ-VC. In the lowest energy region ($\xi_k < \omega_c$), p-p instability dominates over p-h one. Then, U-VC for SC paring interaction plays significant roles in forming Cooper pairs. As a result, we revealed that TSC phase appears near the boundary of orbital- and magnetic-ordered phase. Therefore, microscopic origin of TSC phase comes from cooperation between spin and orbital fluctuations due to the VC beyond RPA+ME theory. In addition, we compared the result of fRG study with that of diagrammatic calculation. Then, we conclude that the dominant contribution comes from AL-type U-VC, which brings mode coupling between spin and orbital degrees of freedom.

Cooperation of el-ph and AFM Fluctuations for SC State

In Chap. 3, we studied two-orbital Hubbard-Holstein model in the presence of B_{1g} el-ph interaction. We revealed that small el-ph coupling g strongly enhances the quadrupole orbital fluctuations due to the significant roles of AL-type χ-VC. In addition, fully gapped s-wave SC phase emerges near magnetic QCP even in the presence of the small g. The required value of g gets smaller as the system approaches to AFM-QCP. Large attractive paring interaction appeared due to the orbital fluctuation U-VC. For this reason, we conclude that microscopic origin of s-wave SC state is the cooperation between AFM fluctuation and el-ph interaction, which is neglected in conventional RPA+ME theory. We stress that U-VC is enlarged only near the Fermi surface. Therefore, local approximation fails to explain large U-VC. We note that retardation effects due to the cutoff energy of el-ph interaction are also important to understand the obtained phase diagram.

S-Wave Superconductivity in $CeCu_2Si_2$

In Chap. 4, we studied two-orbital PAM motivated by recent progress in experimental study on $CeCu_2Si_2$, which revealed that s-wave SC phase appears irrespective of the fact that the system is near AFM phase. To understand the experimental fact, we focus on many body effects due to χ-VC and U-VC as well as strong SOI in f-electron system. In the presence of strong SOI, various higher rank order parameters get to be active. Therefore, higher rank multipole fluctuations induce nontrivial phase transition. Considering multipole fluctuations, we study SC paring mechanism. First, we showed that various magnetic channel fluctuations develop within RPA theory. And also, electric fluctuations are induced by the magnetic fluctuations due to the χ-VC or A_{1g} el-ph interaction. As a result, s-wave SC phase appears near AFM-QCP. The microscopic origin of s-wave SC state is the large attractive paring interaction due to the electric channel multipole fluctuations. We found that the attraction is drastically enhanced by magnetic channel fluctuations through the U-VC. In addition, we found that multi-orbital nature of Fermi surface plays important roles in the proposed mechanism. Especially, $D^{f_1}(0) \approx D^{f_2}(0)$ is needed. We also showed that various types of electric multipole fluctuation including quadrupole and hexadecapole fluctuations work as a attraction for SC paring. In summary, we proposed SC paring mechanism in HF $CeCu_2Si_2$ considering many body effects beyond conventional RPA+ME theory. Essentially, violation of conventional ME theory brought us a new SC pairing mechanism, which is "interference among the different rank of multipole fluctuations".

Multipole Phase

In Chap. 5, we studied 2D Γ_8 quartet PAM to understand the origin of hidden-ordered phase in CeB_6. First, we performed RPA, and showed that various magnetic multipole fluctuations develop. However, magnetic multipole fluctuations cannot explain the experimental facts which revealed that the elastic constant jumps at transition temperature of hidden-ordered phase. Therefore, we go beyond RPA by considering χ-VC for multipole susceptibilities. As a result, we revealed that electric quadrupole O_{xy} fluctuations are induced by the interference between two types of magnetic octupole T_x, T_y fluctuations. In particular, AF-octupole and ferrooctupole fluctuations induced AF-quadrupole phase. Then, we proposed octupole-fluctuation-mediated quadrupole (O_{xy}) phase in CeB_6. Finally, we introduced magnetic field along z-axis and showed that octupole T_{xyz} order is induced by the magnetic field.